A Beginner's Guide to Genealogy 2.0

From the same author

In English

Mastering digital marketing like a boss, Éditions Kawa, with Yann Gourvennec, 2014
Mastering social selling like a boss, Éditions Kawa, with Sylvie Lachkar, 2017
A Beginner's Guide to Genealogy 2.0, Amazon KDP, 2024
50 Ways to Kill Your Business, Amazon KDP, 2025

In French

Médias sociaux et B2B, un mariage d'amour, Éditions Kawa, with François Laurent, 2011
Les médias sociaux expliqués à mon boss, Éditions Kawa, with Yann Gourvennec, 2011
La communication digitale expliquée à mon boss, Éditions Kawa, with Yann Gourvennec, 2013
Ils ont pensé le futur : web social, marketing, e-commerce, Éditions Kawa, Collectif, 2012
Le social selling expliqué à mon boss, Éditions Kawa, with Sylvie Lachkar, 2016
Le digital expliqué à mon boss, Éditions Kawa, with Yann Gourvennec, 2017
Le confinement expliqué à mon boss, Éditions Kawa, Collectif, 2020
Le Guide Pratique de la Généalogie 2.0, Amazon KDP, 2024
Comment Planter sa Boîte en 50 Leçons, Amazon KDP, 2025

A Beginner's Guide to Genealogy 2.0:
Using the Internet, DNA, and Online Databases to Trace Your Family History

Learn to use online databases,
social networks, DNA and digital archives
to successfully build your family tree.

Hervé Kabla

A Beginner's Guide to Genealogy 2.0

Book Summary

A Beginner's Guide to Genealogy 2.0

Foreword

To my family, without whom this book would not have existed...

A Beginner's Guide to Genealogy 2.0

I. Foreword

I discovered genealogy by chance. Like millions of other amateur genealogists, I was looking to learn more about my family's past following the death of one of my grandparents. The family tree was wider than it was deep — concepts of graph theory that we'll discuss later — and I quickly realized that using software would be better than paper. Thus, I started the first branches of a tree that would only continue to grow.

Genealogy is a peculiar activity. Closer to a hobby than a science, it attracts thousands of curious individuals each year, some of whom will engage with it only until they hit their first unsolved inquiries, while others will choose to delve deeper, seeking help from a professional or joining an amateur club.

This book is primarily for those taking their first steps in genealogy. Designed as a travel companion, it aims to guide them through each stage, helping not only to construct their family tree but also to assist others around them in building theirs. And, who knows, maybe even discover distant relatives...

This book is designed for beginners, which is why each chapter is organized progressively. The first chapters are dedicated to the basic principles of constructing a family tree. This is not complicated, but over the past ten years, I have seen so many trees built in convoluted ways that it seemed necessary to share some simple techniques for creating a tree that can be easily used by multiple people over time.

Once these basics are well-established, we'll explore the main genealogy software. The goal is not to endorse any specific software — there are more than a dozen that are quite accessible — but to help the reader understand their fundamental principles, especially in this era of the internet and collaborative approaches.

A Beginner's Guide to Genealogy 2.0

The internet opens new possibilities for genealogists by providing access to online information, making millions of records and administrative documents accessible. Research that previously required risky trips to distant regions or lengthy and tedious exchanges by mail can now be conducted from your computer. However, understanding how it works is crucial.

Finally, we'll discuss the impact of DNA, the risks and tremendous benefits it brings, and the sometimes difficult-to-interpret information it provides. While it is easy to verify paternity or maternity through DNA, it is far more complex to understand what it means to "share 1.6% of DNA" with another individual. An entire chapter is dedicated to this topic.

No matter how far you go in your research, genealogy will remain a fascinating activity. It's best to approach it with every possible advantage, including digital tools.

Happy reading!

H. Kabla
Paris, October 2024

(and don't forget to leave a customer review after reading)

II. Basic Principles of a Family Tree

A. The Family Tree: A Family, Political, and Economic Tool

When talking about genealogy, we immediately think of one expression: "family tree." Representing lineage history in the form of a tree has become the norm. However, this was not always the case, and the very concept of a tree hides particularities and challenges that beginner amateur genealogists may not always know how to handle.

In some passages of the Bible, for example, it is common to trace the genealogy of an important figure not in the form of a tree but as a list or enumeration — a list of ancestors or descendants, depending on the perspective taken. For instance, here is the beginning of the genealogy of Noah's descendants:

> *"Here are the generations of Noah's children, Ham and Japheth; they had children after the flood. The children of Japheth were Gomer, Magog, Madai, Javan, Tubal, Meshech, and Tiras. And the children of Gomer were Ashkenaz, Riphath, and Togarmah. And the children of Javan were Elishah, Tarshish, Kittim, and Dodanim."*

The format of ancient texts didn't always lend itself to a graphic representation of a family tree, and the Bible's very male-centred perspective, listing only male characters, overlooks the presence of women — who are essential to life and make up 50% of humanity... and of our ancestors!

This very patriarchal and linear representation of family history continued to prevail in Antiquity. Egyptian or Assyrian kings, Greek or Persian

11

emperors — almost all referred to a paternal lineage to justify their role and privileges.

It was probably during the Middle Ages, with alliances forged and broken through wars and invasions, that the first trees emerged showing both family branches, going as far back as possible. When two kingdoms united their forces by marrying their children, they had to anchor the alliance in memory, recalling the territories and "houses" linked to each family. The representation of genealogy thus became popular not so much out of pleasure but out of political necessity. A distinguished descendant of a great family could claim special rights in the eyes of peoples subjugated by his ancestors, or among inhabitants of regions under his ancestors' rule.

But while it helped resolve issues of ancestry, the family tree also served to address issues of descent, especially in an era when kingship was a divine right passed from father to son. What to do in the absence of male heirs? What about multiple families or remarriages in cases of widowhood (as divorce was still forbidden by the Church)?

The family tree thus took on a different orientation: previously focused on ancestors, it now became oriented towards the descendants of a particular figure. This inheritance-focused use made the family tree an essential tool for notaries and professionals tasked with resolving inheritance disputes or locating heirs for abandoned property. Given the time needed for such research and the specialization required to achieve tangible results, genealogy gradually became professionalized, with experts in the field able to charge fees based on the results obtained.

B. A Growing Interest

It was only in the second half of the 20th century that genealogy truly developed as a hobby rather than an economic or political activity. With globalization accelerating, people felt the need to reconnect with their roots, to find a heritage, with families dispersed across continents, a rise in blended families, or simply the desire to reunite with lost family members. Genealogy thus flourished in most industrialized countries, where clubs, associations, and magazines dedicated to the topic appeared.

The French Federation of Genealogy, for instance, was established in 1968, currently comprises nearly 150 associations, and claims 60,000 members as of late 2024. It is estimated that by the early 2020s[1], about 10 million French people – who represent 17% of the total population of France - had already interested or engaged in genealogical research or constructed a family tree. The same kind of growing interest can be observed in the US, where a 2021 IPSOS survey[2] found that 33% of Americans said they or a blood relative had researched their ancestry online.

The development of the Internet, of online databases and DNA tests largely contributed to the rapid surge of interest in genealogy, transforming it from a niche hobby into a widely accessible activity. Vast records now digitized and accessible online, genealogical research no longer requires trips to distant archives or browsing through physical documents. Websites offer searchable databases of birth, marriage, and death records, newspapers, and immigration documents, making it possible to trace family history with just a few clicks. Online genealogy software enabled a collaborative approach to genealogy, transforming it from a lonely to a group activity. DNA testing has further revolutionized genealogy by allowing people to connect with distant relatives and uncover their genetic heritage, often revealing information that paper records alone cannot.

---

[1] Source: Geneanet's website
[2] Source https://www.ipsos.com/en-us/majority-americans-think-knowing-their-ancestry-important

C. Drawing a family tree

When you set out to create your family tree, the problem of representation arises. And with human imagination being boundless, you can see many types of family trees, some more or less easy to read and understand. Let's first focus on designing a simple family tree.

Without software, you can start with a pen and a piece of paper, the size of which will depend on the family size. Building your tree involves representing individuals — by their names and first names — and connections between them. The simplest way is to create small boxes or bubbles connected by lines of descent. These will be the "nodes" of our tree.

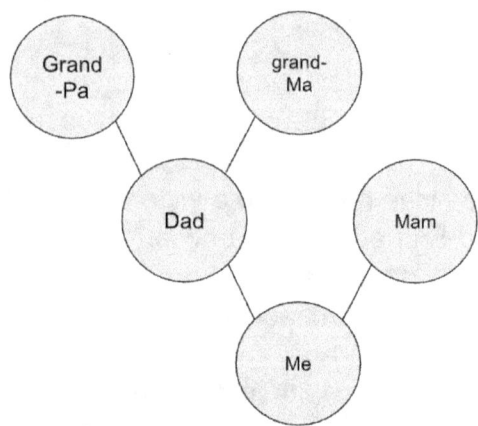

Sample family tree.

However, the way links are distributed differs depending on whether you're creating an ascendant tree to display your parents, grandparents and ancestors, or a descendant tree, to visualize someone's children, grandchildren, and their own descendants.

Basic Principles of a Family Tree

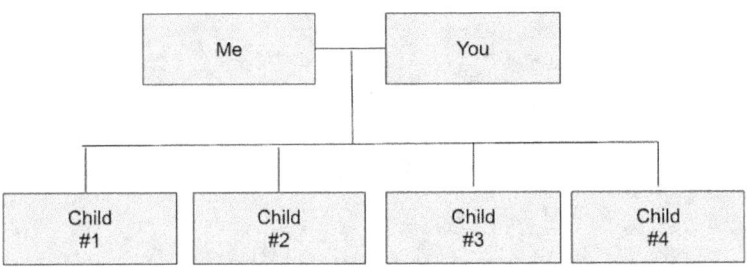

Another sample family tree

Try it yourself, and you'll see it's quite simple. The only challenge you'll face is arranging the nodes of the tree: not too close to one another so you can place different family members, but not too far apart either so it remains easy to read. You can also indicate birth dates and places to keep a record of events.

1. Case of an Ascendant Tree

In the case of an ascendant tree, you simply link each individual to their father and mother. This creates what is known as a "binary" tree, whose size and number of nodes are quite easy to manage, depending on the number of generations represented.

- 1 generation: 1 node (myself alone)
- 2 generations: 1+2 = 3 nodes (myself and my parents)
- 3 generations: 1+2+4 = 7 nodes (myself, my parents, my grandparents)
- 4 generations: 1+2+4+8 = 15 nodes (myself, my parents, my grandparents, my great-grandparents)

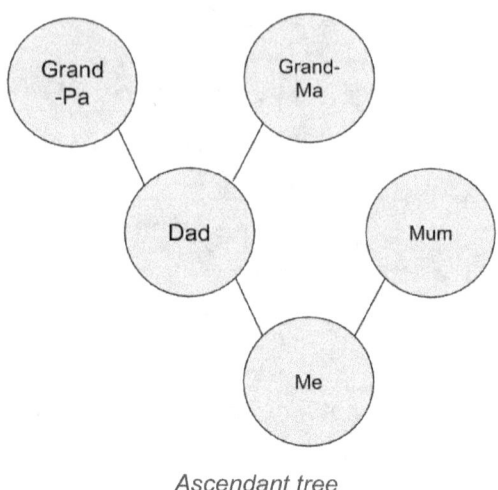

Ascendant tree

And so on. In generation *n*, there are $2^n - 1$ nodes. But this is only in theory. In practice, multiple small issues arise that need to be resolved.

The first issue is that we do not always know all our ancestors. In large families, for example, age disparities among children of the same grandparent create differences in family knowledge. A twenty-year gap in a single sibling group has an immediate impact: the children of the youngest sibling may not know the names of their grandparents, whom they may never have even met. Therefore, you will need to search for these grandparents by questioning uncles and aunts.

Provided, of course, that these relatives are still alive when you create the family tree. This often coincides with the death of one of the parents. If the deceased was the last child of their family, they may have no surviving siblings (your uncles and aunts), creating an "artificial" horizon on the family tree. Rest assured; it is often possible to trace back to the grandparents through civil records.

Basic Principles of a Family Tree

The second issue is that, several decades ago, it was still common to marry cousins. Modern transportation was not widespread, and people often spent their lives in the same town or village, associating only with the same individuals, often within their own family. Consanguineous marriages were still common in the early 20th century. While this had implications for the development of hereditary diseases, it also impacts your family tree. If your parents were cousins, it means that your grandparents on both sides were siblings, sharing the same parents — your great-grandparents. You will end up with duplicate information. This isn't too problematic on paper and only becomes noticeable after three generations. But as we'll see later, this presents challenges on a computer.

2. Case of a Descendant Tree

The representation of a descendant tree follows the same principles of nodes and links, but with a slight difference compared to an ascendant tree. If you want to represent all the links between you and your children, there is no issue. However, as soon as you wish to include spouses — which seems quite logical — you will end up doubling the connections. If you also have a blended family, it quickly turns into a tangled mess that is hard to untangle.

To represent a descendant tree with spouses, we slightly modify the attachment points between nodes. Instead of linking individuals directly, we first link the parents to each other, then link each child to the midpoint of the connection between their parents.

A Beginner's Guide to Genealogy 2.0

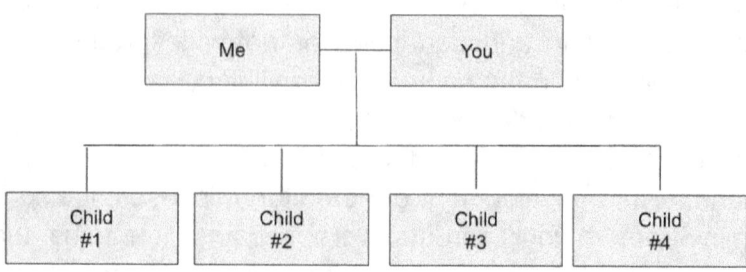

Descendant Tree

This approach makes such a descendant tree more readable, as each descendant is linked to a single element: the connection between their parents.

For readers interested in the algorithmic representation of a descendant tree, this effectively introduces a new type of node, the "family" node. Each "individual" node is connected to an ascending "family" node (its parents) and to multiple descending "family" nodes (one per couple, whether they have children or not).

The representation of a descendant tree does not pose the same problems as an ascendant tree, since when establishing your tree, you are generally aware of your children's offspring. However, theoretically, you may encounter the same issue with marriages between cousins.

In terms of size, we do not follow the same rule based on powers of 2 for a simple reason: the number of grandchildren is not the same for each child. This results in what is known in computer science as an unbalanced tree, and when "drawing" this tree, you need to account for each child's number of descendants. A good tip in this case: choose a landscape format rather than portrait by rotating your paper by a quarter turn.

3. Global Family Trees

Once you have enjoyed building your ascendant and descendant trees, you may be tempted to take the exercise a bit further and represent your cousins and their offspring. This essentially means creating descendant trees stemming from the nodes of the ascendant tree...

Nothing really too complicated, except in this case... the size of the paper. Don't hesitate to use multiple sheets if your tree extends beyond the edges. You'll just need to keep track of the position of each sheet, perhaps by adopting a numbering system, such as assigning row and column numbers to each sheet in the final layout.

D. Data and Metadata

We mentioned earlier that when creating the nodes of a tree, you should include the first and last name of everyone. This is correct but incomplete. A tree, whether explicitly or implicitly, aims to keep an accurate record of each ancestor, so it's common to add some additional information, often referred to as "metadata":

- When known, include the date and place of birth, and if the person has passed away, the date and place of death.
- Indicate gender, especially for names that may be ambiguous.
- For women, you may also mention the maiden name as well as the married name.
- In cases of marriage or divorce, include the date and place of marriage, and possibly the date of separation.
- If available, it's also nice to associate a photo of the person, which can be pasted or taped onto the tree represented on a large sheet of paper. If using software, you'll need photos in digital format: easy for recent photos, and a bit more complex for

older photos that may need to be scanned or photographed with a smartphone.

When working on trees with several thousand nodes, it's possible to encounter multiple individuals with the same first and last names. If dates are approximate or information is incomplete, distinguishing between them can be difficult. This is why, whenever possible, you should add details to remove ambiguity, such as occupation or educational background. In terms of occupation, the same person may have held different jobs or worked in different cities at different times. Be as precise as possible in these cases, and use software to facilitate the work and prevent your tree from becoming unreadable...

E. Additional Uses of Genealogy

If family trees have become quite widespread, it's also because their uses have expanded beyond the family circle. Let's go through some of them, from the most essential to the most intriguing...

1. Education

In some countries, creating a family tree is a required activity for all students during elementary school. The goal is both to reconnect each child with their family, their grandparents, and to demonstrate, through shared trees, the diversity of backgrounds of the individuals living in that country.

From an educational perspective, such a project offers numerous advantages. Through the creation of their family tree, children can discover not only family history but also the history of their own country, including various waves of immigration and exoduses. Building a family tree requires both rigor and focus on a subject that is relatively easy to access. It develops the child's critical thinking by encouraging them to compare multiple contradictory leads. It also allows the integration of skills from various disciplines: history, geography, mathematics (to calculate ages or introduce the concept of averages), and even biology.

The creation of a family tree at a young age is also a factor in building cultural identity and developing self-esteem through the connection to parents and ancestors. Furthermore, it can become a factor in fostering community engagement or, conversely, national identity.

2. Heir Research

The search for heirs of unclaimed estates, as previously mentioned, is one of the common applications of genealogy. Migration flows resulting from upheavals or tragedies of the last century have made the search for ancestors and descendants a significant use, sometimes leading to unexpected consequences when an unforeseen heir comes forward.

Many professional genealogists occasionally rely on family trees built by individuals for personal use. Using a death date, for instance, they can "match" profiles with trees created on online genealogy software like MyHeritage or Geni and contact their creators. It is then up to the latter to decide whether to respond to the inquiries from professional agents, who usually work for compensation on behalf of notarial offices.

3. Medical Research

It is not so much genealogy as the development of genetic biology that presents medical interest. The presence of a mutation associated with a risk of a rare disease in an individual may prompt a search for family members who share the same mutation. Knowing one's ancestors and their descendants thus allows anticipation of risk and enables more frequent monitoring where necessary.

The procedure involves, for example, searching for kinship links between individuals affected by rare diseases to understand the mode of transmission (whether through the father or the mother, for instance). Most of the time, these rare diseases arise from the mutation of one or more genes. When only one parent carries the mutation, the child does not develop the disease but may pass the same mutation to their own children. However, when both parents carry the same mutation, the child develops the condition.

To anticipate risks within highly endogamous populations, where consanguineous marriages were common, certain regions have implemented significant medical measures. For example, in the Grand-Ouest region of France, the GEM-EXCELL[3] network integrates several university hospitals in Brittany, combining research, training, and public communication initiatives.

4. Criminal Investigation

Solving criminal investigations through genealogical research has seen rapid development in recent years, mainly due to advances in DNA analysis techniques. By analysing DNA left by a criminal and comparing it with large DNA databases, the search for suspects can be accelerated, and some cases have recently been solved this way[4]. This also partly explains why DNA sequencing by the public is prohibited in some countries, such as France.

The typical procedure is as follows: investigators collect DNA fragments from a crime scene to establish the "genetic profile" of the person being sought. This DNA is then compared with large DNA databases, such as GEDmatch or FamilyTreeDNA, to find familial matches (refer to section V.B.5 for a more detailed explanation and the DNA match table).

---

[3] https://www.gem-excell.fr/
[4] For example, see the Thomas Martin Elliot case
https://www.nbcnews.com/news/us-news/50-year-old-colorado-cold-case-solved-womans-killer-identified-dna-tec-rcna136805

Investigators trace back to potential common ancestors and then map out all family trees descending from those ancestors. This process produces a cohort of potential suspects whose DNA is compared to the initial profile. Through these comparisons, the search can narrow down to a specific family branch, ultimately leading to the identification of the person sought.

This fascinating technique has been used to solve cold cases where DNA evidence - such as a strand of hair, a drop of blood, or semen - was preserved. However, it raises ethical questions about the consent of families involved in the investigation and the respect for privacy.

5. Statistical Analysis

Large family trees, containing several hundred thousand individuals, can also be used for statistical purposes. It's possible to work on the digital representation of such trees to identify the most common first names, the average family size, the demographic evolution of a surname, or any type of inquiry with historical or sociological relevance. If you know how to handle such tools, which requires some programming skills, the only limit is your imagination...

Libraries for computer programming, especially in Python,[5] exist to work with GEDCOM files exported from family trees (for more on the GEDCOM format, see section III.E). If you know how to handle such tools, which requires some programming skills, the only limitation is your imagination.

To spark some ideas, here are a few intriguing possibilities:

- Historical Demography: You could calculate the average age at marriage, the age of parents at the birth of their first child, or average life expectancy. It is also possible to analyze how family

---

[5] See for example: https://pypi.org/project/python-gedcom/

sizes have evolved across generations and compare this evolution by birthplace (city, country).

- Geographical Mobility: You can examine migrations between various locations—cities, regions, countries—and create maps to visualize family movements over time.
- Socio-Economic Trends: Study shifts in professions across generations (requiring specific metadata collection) or analyze the correlation between education level and family size.
- Health and Longevity: Identify trends in life expectancy or, by noting causes of death, explore the prevalence of certain hereditary diseases.

After data extraction, traditional statistical methods (cohort analysis, multivariate analysis) are applied. These datasets can also be used to create innovative visualizations, such as infographics or timelines.

III. Genealogy software

A. Why Use a Genealogy Software?

As we've seen earlier, it is entirely possible to begin personal genealogical research using tools as simple as a pencil and a few sheets of paper. However, if this work is to take on a certain scale, it becomes much easier to rely on specialized software. Before reviewing the most common software, let's start by presenting the main benefits offered by such programs.

1. Dealing with Trees of Unlimited Size

As mentioned earlier, once the family tree, whether ascendant or descendant, starts to fill up, you quickly become limited by the size of the paper used. Even if you work on multiple sheets placed end-to-end, let's admit, it's not very practical.

On genealogy software, you work on a virtual sheet with unlimited dimensions. Of course, as your tree grows, it becomes increasingly difficult to display it entirely on the screen — even with a television screen — and you must settle for showing only a partial view. Most software, like MyHeritage, allows you to specify how many generations you wish to display at once: 1, 2, 3, up to 7, for example.

2. Searching for People

When you have a family tree with a few hundred or even thousands of individuals, it becomes difficult to remember each branch and each person. To locate a specific person, genealogy software typically offers a "search" function, allowing you to find all individuals whose first or last name begins with the letters you type on the screen.

25

3. Printing Your Tree

With software, you can print the family tree as many times as you want, or print a part of the tree, which you can share with family members interested in their ancestors or descendants.

4. Storing Various Kind of Data

As mentioned earlier, with a paper-based tree, it quickly becomes challenging to store detailed personal data, such as birth and death dates, cities, occupations, addresses, etc. Software includes specific fields for such information.

5. Storing Digital Photos

Among the data associated with each individual, photos are often the most sought after: what did grandfather or grandmother look like? Who are the cousins that resemble each other most or, on the contrary, have no shared traits?

With genealogy software, your family tree can easily double as a family photo album. And since we now mostly use digital images, it's much easier to manage compared to printing and pasting small pictures on paper, as children often do when asked to create a family tree in elementary school.

B. Software to Install

The first genealogy software appeared in the 1980s, before the existence of the Internet. These programs had to be installed on personal computers, usually PCs running Windows. Some of these programs have disappeared, but others still exist and are perfectly suitable for personal use.

Most of these programs are paid software, developed by publishers whose main source of income is these sales. Paid software generally offers a free trial with a limited-sized tree of a few dozen individuals, which is sufficient to test the interface and check if all desired features are present.

With the development of open-source software, fully free genealogy programs have emerged, with no limit on tree size and increasingly advanced features.

In the following paragraphs, we'll take a quick look at the main programs you could use to create your tree. For a detailed review of all genealogy software, free or paid, I recommend consulting the Wikipedia page[6] dedicated to them: you'll find information on the latest updates, version numbers, and supported operating systems (Microsoft Windows or Mac OS).

1. Paid Software

Remember that most of these programs offer free access up to a certain tree size.

a) Ancestral Quest

Ancestral Quest (https://www.ancquest.com/) is likely the first software capable of accessing and synchronizing its data with FamilySearch, which we will discuss later. Launched in 1994 and initially developed for Microsoft Windows, it has, in recent years, a version designed for MacOS users.

---

[6] See https://en.wikipedia.org/wiki/Comparison_of_genealogy_software

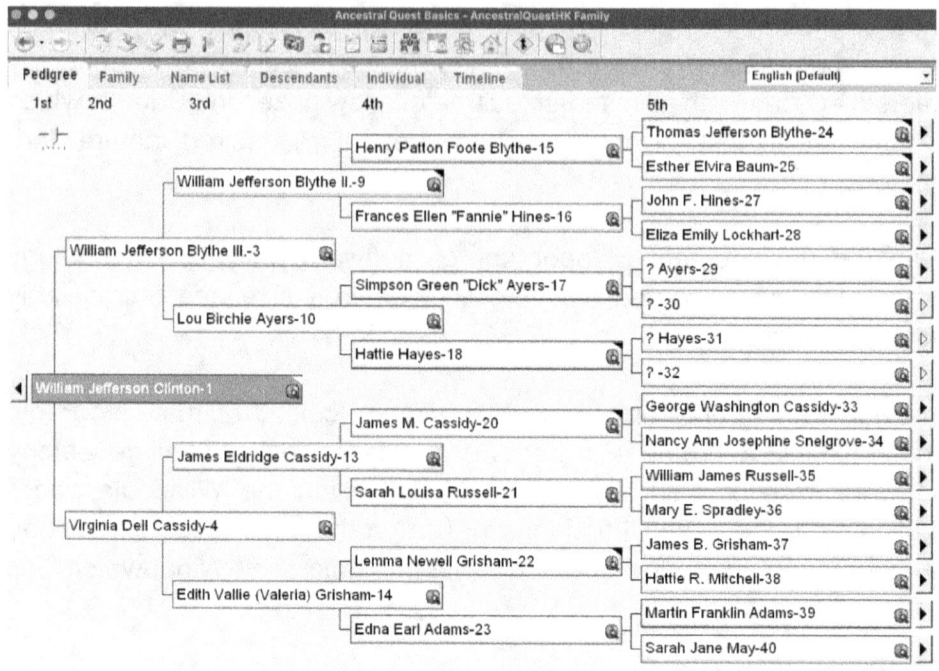

AncestralQuest

b) FamilyTreeMaker

FamilyTreeMaker (https://www.mackiev.com/) is probably the oldest paid genealogy software still available for purchase. Its first version dates back to… 1989. Developed by Kenneth Hess for Banner Blue Software, it was later distributed by several different companies, through successive acquisitions by SoftKey and Mattel, eventually ending up with Ancestry.com, an online genealogy software publisher.

In 2015, this company announced it would stop supporting FamilyTreeMaker, prompting protests from its many users, leading to the software's acquisition by MacKiev. FamilyTreeMaker works on both Microsoft Windows and MacOS.

Genealogy software

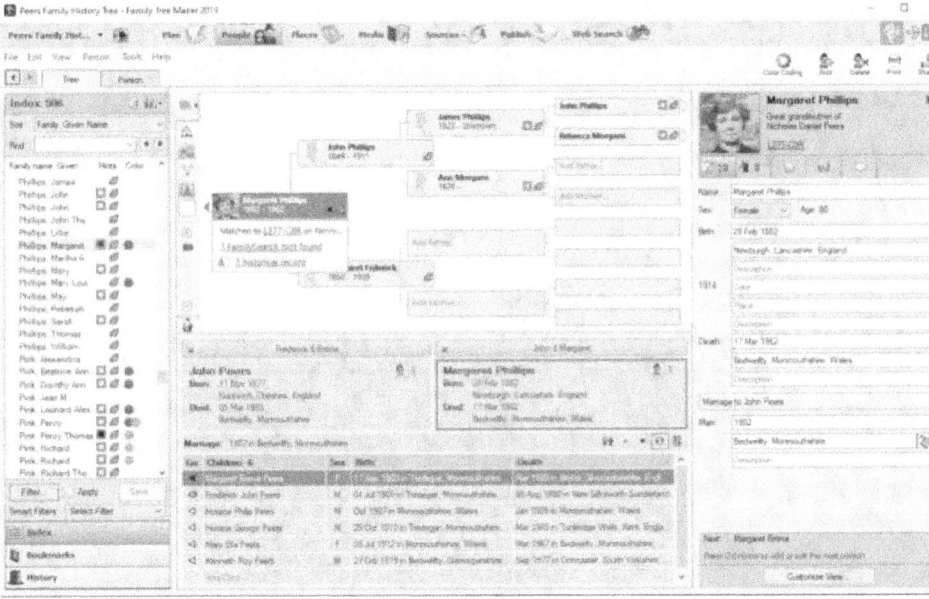

FamilyTreeMaker

c) GenoPro

GenoPro (https://genopro.com/) is a genealogy software that I hold dear personally, as it was the one I started with in the late 90s. With a very simple interface, making it ideal for beginners, it unfortunately only works on PCs under Microsoft Windows. GenoPro supports the GEDCOM format and offers a unique feature: creating genograms, which allow for representing all events concerning a given individual on a single graphic.

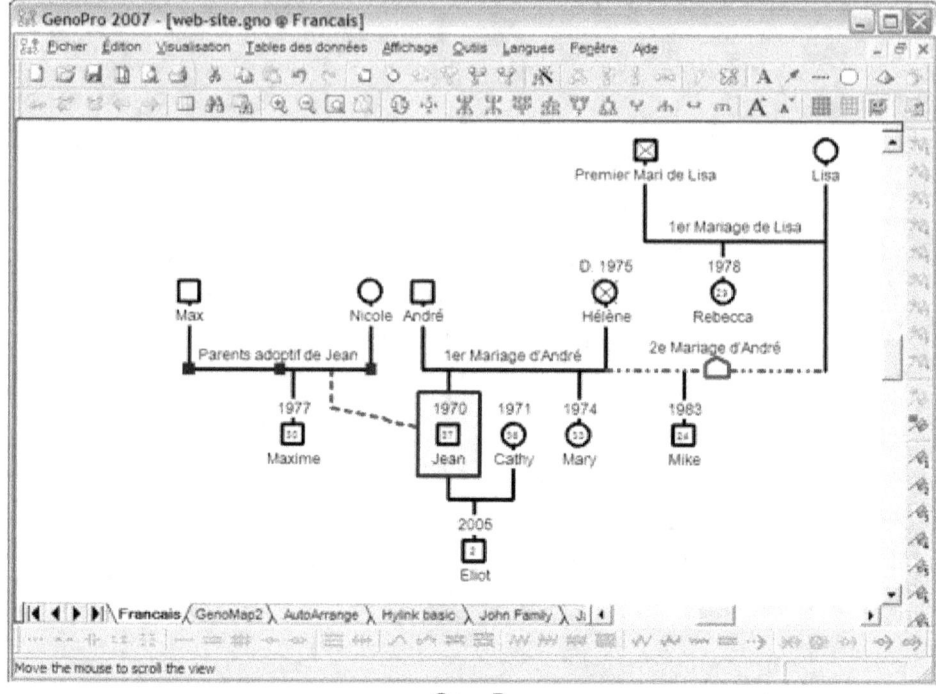

GenoPro

2. Free Software

Here are some free programs you can install on your computer to get started on your family tree.

a) Ancestris

Developed in the early 2000s and maintained by a small team of genealogy enthusiasts in France, Ancestris is free software offering a broad range of features, so you won't need to try another. Backed by a non-profit association, the software is intended to remain free for life, with no limit on the amount of data you can enter.

Ancestris works on Microsoft Windows, Linux, and MacOS. You can download the different versions and flavors from the Ancestris website: https://www.ancestris.org/.

b) Gramps

Gramps (short for *Genealogical Research and Analysis Management Programming System*) is an open-source software, meaning its code is freely available for anyone who wishes to contribute to the project.

You can download it from its official page: https://gramps-project.org/. Developed in Python, Gramps runs on multiple operating systems — choose the one that suits you (probably Microsoft Windows, MacOS, or Linux).

Gramps supports the GEDCOM format, which we'll discuss further, allowing you to import family trees from other programs without any issues or size limitations.

Finally, Gramps is a comprehensive program with a community of Add-On developers who create extensions to enhance its features. However, it's not the easiest to use, and you may be put off by its somewhat austere interface.

C. Online Software

Since the early 2000s, standard software installed on personal computers has gradually been replaced by software accessible online, offering greater flexibility, the ability to use it from different computers, and frequent, hassle-free updates.

Online software is accessed through a website, such as www.myheritage.com for the software provided by MyHeritage. These sites are generally available in multiple languages.

Online software offers several advantages, along with a few drawbacks. The main benefits include:

- No need to be a computer expert to install, configure the license, or update the software.
- Access is possible even without your own computer, or from a smartphone, although it may not be as visually comfortable.
- Features evolve over time without needing to reinstall the software. For example, they might offer access to civil registry records.
- Data is stored online, in databases owned by the software publisher rather than on your personal computer. The software publisher can then compare this data and identify matches or common individuals, allowing you to find common ancestors and connect with family members.
- Most importantly, you can — depending on your subscription and permissions — access family trees uploaded by other users, allowing you to compare with your own information and correct certain data, such as dates, places, or the spelling of names.

The main drawbacks include:

- Most are paid software with a subscription model, requiring an annual fee, which can become costly over time.
- Your data is not stored locally but with the publisher, which may present security issues in case of a site hack.
- If you stop paying your subscription, you face the sensitive question: what happens to your data?

a) Geneanet

Geneanet (www.geneanet.org) is a French genealogy site created in 1996 by Jacques Le Marois, Jérôme Abela, and Julien Cassaigne. Independent until 2021, Geneanet was acquired by the American group

Genealogy software

Ancestry. To date, Geneanet hosts around 1 million family trees, representing several billion individual records.

Geneanet includes the standard features of an online genealogy platform. Additionally, it has multi-criteria search engine that provides access to public trees and several death record databases for subscribed members.

Here is an example of a family tree representation with Geneanet. The level of detail depends on the zoom level used in the viewing window.

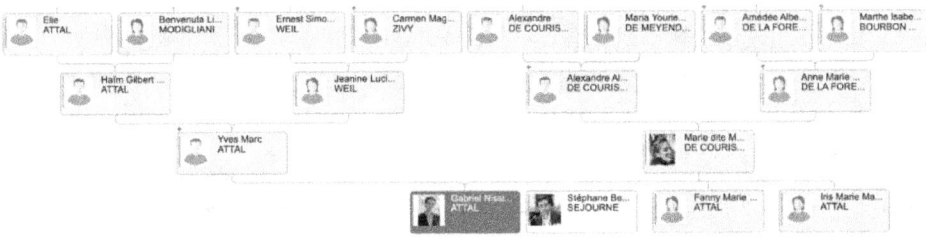

In addition to the graphical representation, which is not always easy to use, Geneanet also offers a text-based view, which is extremely practical for navigating from one family to another.

♂ **Gabriel Nissim ATTAL**

· Né le 16 mars 1989 - Clamart, 92023, Hauts-de-Seine, Ile-de-France, France
· Âge : 35 ans
· Homme d'État Français, Premier Ministre (depuis 09/01/2024)

🖼 3 médias disponibles

Parents

· Yves Marc ATTAL 1948-2015
· Marie dite Marika DE COURISS 1964

Union(s)

· PACS le 6 janvier 2017, Paris 3ème, Ville de Paris, Ile de France, France, avec Stéphane Benoit Lionel SEJOURNE 1985 , séparés

Fratrie

· ♂ **Gabriel Nissim ATTAL** 1989
· ♀ Fanny Marie Madeleine ATTAL 1990
· ♀ Iris Marie Madeleine ATTAL 1992

Demi-frères et demi-sœurs

Du côté de Yves Marc ATTAL 1948-2015
· avec Véronique Marguerite Madeleine BLANCHARD ca 1950
  · ♀ Noémie Madeleine ATTAL 1975

b) Geni

Geni (www.geni.com) is a California-based genealogy site created in 2006 by David Sacks, Alan Braverman, and Amos Elliston. The founders' original idea was to create a single family tree collaboratively developed by all its users. Thus, instead of having one tree per user, the site would manage only one tree, with each user responsible for updating the part that concerns them. Each user can only view data related to their own family (ancestors, descendants, relatives), or data from portions of the tree that another user has granted them access to.

This approach sets Geni apart from all other online software, which have a more traditional approach where each user manages their own tree, and the same individual can appear in multiple separate trees. Geni's approach is more "intelligent" because it allows for the correction of potential errors once and for all, whereas in traditional software, the same person may appear with slightly different data, potentially leading to cascading errors.

Genealogy software

In 2012, MyHeritage (see below) acquired Geni as part of its acquisition strategy. Geni tree data is therefore accessible from MyHeritage.

Here is an example of a tree on Geni. You can see small green clouds containing numbers, which help avoid cluttering the tree representation and allow access to tree portions linked to the profile marked by that icon.

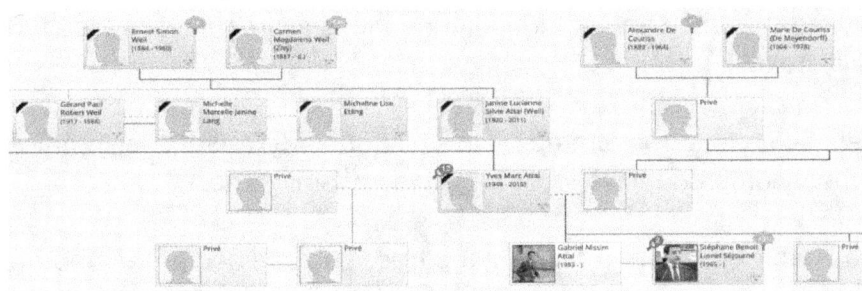

c) MyHeritage

MyHeritage (www.myheritage.fr) was founded in 2003 by Gilad Japhet from a kibbutz in Israel, where the company continues to operate. It is a genealogy platform accessible via a web browser and a mobile app compatible with both iPhone and Android. In 2024, the company claimed over 6 billion individual records and around 50 million online family trees.

MyHeritage has pursued a growth strategy typical of a tech start-up. After a funding round in 2008, it acquired competitors such as DynasTree, FamilyLink, and more recently, Geni, as well as the French platform Filae.

MyHeritage offers various tools to its users, depending on the type of subscription chosen (the most comprehensive subscription costs around 250 euros per year). These tools include several national death record databases, archive databases like American port archives, newspaper archives, and a search engine allowing multi-criteria searches (first name, last name, year of death or birth, parents' names, etc.) across all

MyHeritage databases. Finally, MyHeritage provides an impressive automatic "match" feature, capable of identifying identical individuals whose names or first names were transcribed in different languages.

In 2016, MyHeritage ventured into DNA sequencing, enabling users to discover family links based on the percentage of shared DNA between two users. Since 2023, however, MyHeritage has stopped shipping its DNA kits to France and other countries with strict regulations (see further details below).

Given all the features offered by this platform, MyHeritage is, in my view, the most comprehensive tool for building a family tree.

Here is an example of a MyHeritage tree. You can see green and brown markers. The green markers allow a match with other MyHeritage trees. The brown markers allow a match with a database (for example, the public figure directory on WikiData) or with trees from other platforms (Filae, Geni).

Here is an example of a MyHeritage tree. You can see green and brown markers. The green markers allow a match with other MyHeritage trees. The brown markers allow a match with a database (for example, the public figure directory on WikiData) or with trees from other platforms (Filae, Geni).

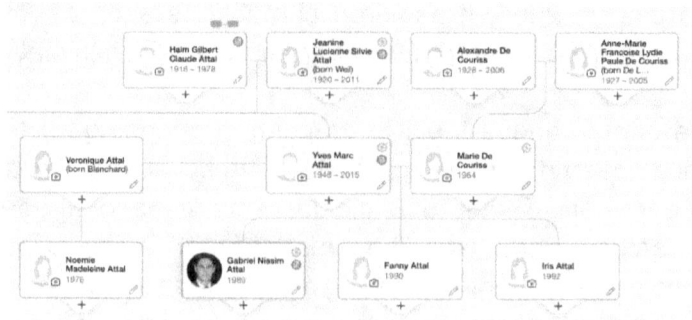

MyHeritage offers several modes for representing family trees, including an impressive fan view that allows users to visualize ancestors over 6 or 7 generations in a synthetic way.

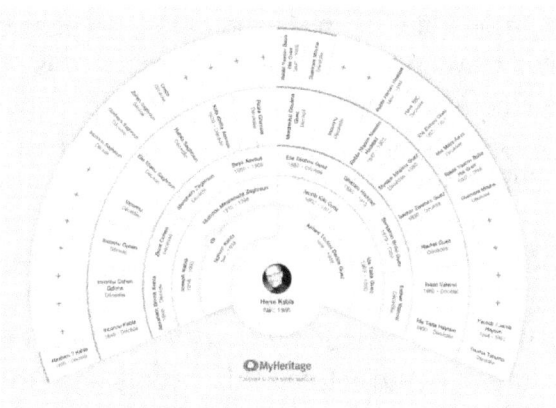

D. Common Mistakes and How to Correct Them

You will quickly realize the usefulness and ease of genealogy software. They allow you to do everything you could with a pencil and paper, and much more. They are packed with sorting, searching, and data storage features, often more ingenious than the last.

However, mistakes do happen, regardless of one's familiarity and expertise with the subject. This is especially evident when using online software and accessing trees created by users who are less meticulous.

Here is a quick list of the most common mistakes you may encounter or make out of oversight or lack of knowledge.

These mistakes are easy to correct, so don't hesitate to fix them as soon as possible: a forgotten error eventually spreads and tarnishes the rest of your genealogy work...

- Reversing the father and mother: After several minutes of entering names, first names, and dates, it's possible to accidentally switch first names. This is very easy to correct.
- Entering incomplete dates: Similarly, due to fatigue or a worn-out keyboard, you might miss a digit in a date format, leading to strange dates, sometimes as early as the 10th or even the 1st century. Remember to enter the complete year, such as 1975 rather than just 75.
- Confusing descendant and ascendant trees: This is a subtler mistake, sometimes made by users who haven't fully understood the structure of a family tree. This error is easy to spot: an ascendant tree for each individual is a binary tree (two parents per individual), while a descendant tree can have multiple branches. For a refresher on these basics, see the chapter on Basic Principles of a Family Tree.
- Connecting an individual in the wrong place: This is also a classic mistake, made either by accident or due to lack of precise information. It's usually noticed through date inconsistencies. For example, a child's birth date might appear two or three years before their father's, indicating it's likely their sibling instead.
- Creating duplicates: This classic mistake is often due to the presence of a cousin couple: their grandparents are the same, and instead of linking the sibling parents to the same grandparents, the grandparents are duplicated, complicating the tree. Often, users who have created duplicates notice the problem but don't know how to fix it, choosing to fill in the ascendant tree on only one side to avoid entering the same ancestry twice.

E. Data Exchange Formats and Data Migration

You may start exploring genealogy by creating your family tree with one software, but eventually feel limited by it due to a lack of features or because it becomes too complex to use. In such cases, you might want to switch to another software to manage your tree. Refer to the following sections to help with your choice.

However, it would be a shame for the work done on the first software to be lost, forcing you to start from scratch with the new software. Don't worry; others have encountered this problem before you and have provided an efficient, surprisingly effective solution...

This solution is called GEDCOM, an acronym for *GEnealogical Data COMmunication*. It's a file format, much like JPEG or HTML formats. This format was invented in the 1980s by members of The Church of Jesus Christ of Latter-day Saints, whose members are commonly referred to as "Mormons." Strongly present in Utah (particularly in their capital, Salt Lake City), they place special importance on sacraments for their ancestors and have thus engaged in extensive genealogy research over several decades.

In France, for example, the Mormons have made agreements with the several national archives organizations in the world to digitize millions of civil registry records existing in paper format. All the data collected by the Mormons is accessible online and free of charge on the site www.familysearch.org.

Being among the first users of genealogy software, members of this Church were the first to face data exchange challenges when comparing multiple trees, for example. They specified and created the first format enabling interoperability between genealogy software, the GEDCOM format.

On the next page, you can see the first few lines of a sample GEDCOM file. Such a file can easily have tens of thousands of lines, depending on the size of the tree you manage, and the quantity of metadata you have put inside.

What's important to know is that most software allows you to export the data you've entered in GEDCOM format, as well as import data from a GEDCOM file.

So, you can easily switch software by following these steps:

- Log in to the software you want to leave.
- Save your tree in GEDCOM format (a file with a .ged extension).
- Close the software you're leaving.
- Open the new software you'll be using.
- Import the .ged file you just created.
- Check that everything has been imported correctly and correct any import issues (such as date formats or accents).

You don't need to learn how to decode this file format. It's actually quite indigestible, designed for software rather than humans.

As you'll see, importing and exporting data via GEDCOM is quite simple. But be aware of certain limitations.

- Ensure that your subscription (or license) on the new software allows you to manage a tree of the same or larger size; otherwise, you won't be able to import your entire tree.
- You might be tempted to import a relative's tree into your own, to merge data from both families, for instance by overlapping descendants of a common ancestor. This might seem like an interesting feature, but unfortunately, genealogy software doesn't yet offer this functionality. They simply create a second

Genealogy software

tree as a separate window, without merging data from your two trees.

```
0 HEAD
1 GEDC
2 VERS 5.5.1
2 FORM LINEAGE-LINKED
1 CHAR UTF-8
1 LANG English
1 SOUR MYHERITAGE
2 NAME MyHeritage Family Tree Builder
2 VERS 5.5.1
2 _RTLSAVE RTL
2 CORP MyHeritage.com
1 DEST MYHERITAGE
1 DATE 10 JUL 2024
1 FILE Exported by MyHeritage.com from Famille Kabla on Wed, 10 Jul 2024
1 _PROJECT_GUID 5BA49B3A60F537AD3FA163E098243CF4
1 _EXPORTED_FROM_SITE_ID 552618321
0 @I1@ INDI
1 _UPD 8 DEC 2021 02:46:19 GMT -0500
1 NAME Joseph /Kabla/
2 GIVN Joseph
2 SURN Kabla
1 SEX M
1 BIRT
2 DATE 20 FEB 1916
2 PLAC Tataouine, Tunisie
1 DEAT Y
2 DATE 16 DEC 1993
2 PLAC Paris, France
2 NOTE Cause: Medical
2 AGE 77
1 FAMS @F2@
1 FAMC @F1@
1 SOUR @S500013@
2 PAGE https://www.myheritage.fr/research/collection-40000/geni-world-family-
tree?itemId=416364097&action=showRecord
2 QUAY 4
2 DATA
```

A Beginner's Guide to Genealogy 2.0

IV. Databases

A. From Paper to Digital

It would be a mistake to think that the main contribution of computers to genealogical practices is limited to the use of software for simple tree representations. The changes brought about by digital technology affect other aspects, notably the storage of data essential for genealogical research, such as civil records.

A significant part of the work, when building a family tree, involves researching ancestors by consulting archives maintained by various "authorities": the administration (e.g., civil records, naturalization decrees), religious institutions (e.g., religious marriage records kept in synagogues), national libraries, immigration records, the American tax fraud database, etc.

Whereas previously people had to travel to consult these types of documents, it is now possible to conduct such research without traveling, using a computer. A massive digitization effort has been launched in recent years, not only by the Mormons, whose project we mentioned in the previous chapter, but also by various administrations in many countries.

This chapter offers an overview of these vast databases, which you need to learn to navigate to extract useful information. Without aiming for exhaustiveness, which would be of little interest, we'll look at how to get by with some tools of varying accessibility so that you can understand the philosophy behind this type of research.

B. National Archives

National archives can be considered as the memory of a country. Usually designed as large vaults to keep in good condition vast amounts of printed documents, ranging from peace treaties to original copies of national great writers, including copies of the major newspapers printed in the twentieth century. For decades, these institutions offered access to a limited number of documents to registered visitors. But since the 2000's, due to large progress in the process of digitization, access to more and more documents is proposed through the national archive's websites.

As an apprentice genealogist, the large number of archives may not be so useful, unless you are the descendant of a famous president or diplomat. However, these organizations usually dedicate specific parts of their websites to genealogists looking for specific data, like naturalisation files.

Of course, each country has its national archives websites, and there is no unique entry point for all of them. Depending on the set of services offered by the website, you will be able to browse online collections of civil registry documents like birth certificates or naturalization papers dating from more than a couple of decades.

To help you start you own research, here is a list of national archives websites and their genealogist's section for some western countries.

Databases

1. United States

The US national archives website is located at https://www.archives.gov/. You can find several sections like veterans record or America's founding documents. There is a section dedicated to genealogists and located at https://www.archives.gov/research/genealogy.

Among the most interesting section, you should explore the data provided in the census, naturalization, or immigration records section.

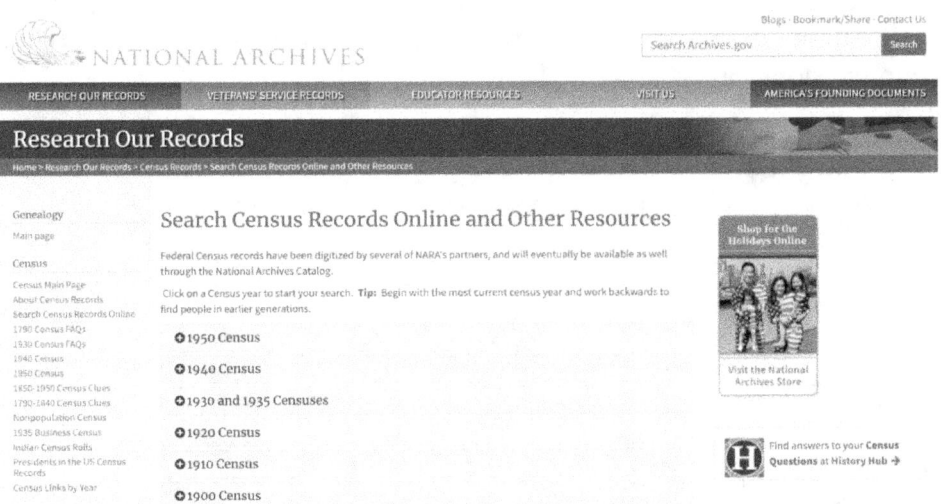

This sites also gives access to a very powerful tool, the AAD (Access to Archival Database) website, located at https://aad.archives.gov/aad/. This site is a kind of single entry point to several databases, organized by categories: time spans (1800-1900, 1900-1940, 1940-1955, 1955-1965, 1965-1975, 1975-1985, 1985-1995 and 1995-2005), military personal, immigrants, war casualties, etc.

Each category provides access to several databases: corporate information, American buildings, Numident, military assistance program, war casualties, grants and insurance loans, etc. Numident is probably the most useful ones for genealogy searches, although you could get additional data from the other databases.

Numident stands for "Numerical Identification System". It is the Social Security Administration's computer database file of an abstract of the information contained in an application for a United States Social Security number (Form SS-5). It contains the name of the applicant, place and date of birth, and other information. The Numident file contains all Social Security numbers since they first were issued in the US, in 1936.

The database is split in several chunks, organized by the first letter of the last name of each individual. Hence, beware to start your search request from the appropriate one.

Numerical Identification Files (NUMIDENT), created, 1936 - 2007, documenting the period 1936 - 2007 - *Record Group 47*	search	146,870,508
Application (SS-5) Files, 1936 - 2007 (*Last Names A through B*)	search	8,962,060
Application (SS-5) Files, 1936 - 2007 (*Last Names C through D*)	search	8,837,364
Application (SS-5) Files, 1936 - 2007 (*Last Names E through G*)	search	7,614,534
Application (SS-5) Files, 1936 - 2007 (*Last Names H through J*)	search	7,795,631
Application (SS-5) Files, 1936 - 2007 (*Last Names K through L*)	search	5,998,506
Application (SS-5) Files, 1936 - 2007 (*Last Names M through N*)	search	8,093,260
Application (SS-5) Files, 1936 - 2007 (*Last Names O through R*)	search	8,610,386
Application (SS-5) Files, 1936 - 2007 (*Last Names S through T*)	search	9,738,423
Application (SS-5) Files, 1936 - 2007 (*Last Names U through Z and non-alphabetic*)	search	6,530,565

Once you have selected a group, the website displays a very simple form where you just enter bits of information you already know (for example, last name or year or birth), and the engine provides you with all the records that match your query.

Databases

Fielded Search

File unit: Application (SS-5) Files, 1936 - 2007 *(Last Names S through T)*
in the Series: Numerical Identification Files (NUMIDENT), created 1936 - 2007, documenting the period 1936 - 2007 - *Record Group 47* (info)

ℹ️ These files do not contain records of all Social Security Number applications. The files only contain applications of deceased individuals. You may wish to View the FAQs for this series.

Search this file

[   ] Search Advanced Search

Enter values below to search within fields.   ⊕ show more fields

Field Title		Enter Values
SOCIAL SECURITY NUMBER	with all of the values ⌄ [ ]	Sample Values
FIRST NAME	with all of the values ⌄ [ ]	Sample Values
MIDDLE NAME	with all of the values ⌄ [ ]	Sample Values
LAST NAME	with all of the values ⌄ [ ]	Sample Values
DATE OF BIRTH (MONTH)	Select from Code List	
DATE OF BIRTH (DAY)	Select from Code List	
DATE OF BIRTH (YEAR)	equals ⌄ [ ]	Sample Values
PLACE OF BIRTH CITY	with all of the values ⌄ [ ]	Sample Values
STATE OR FOREIGN COUNTRY OF BIRTH	Select from Code List	

2. United Kingdom

The United Kingdom national archives website is located at the following url: https://www.nationalarchives.gov.uk/.

Civil registry documents must be ordered, the process can be done online from this website: https://www.gov.uk/general-register-office. Ordering is not free.

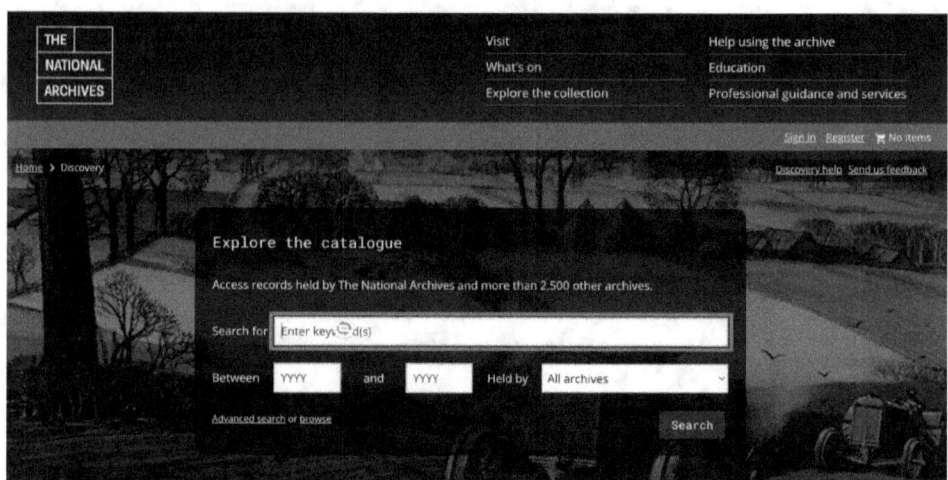

3. Italy

The tradition of archiving civil records in Italy dates to the introduction of Napoleonic civil registers, in 1806., in some regions annexed to the French empire. Before that, registers were maintained by religious institutions. Catholic parish registers were the main source of data. In Livorno (Leghorn), where a very active Jewish community was settled, religious registers date back to the beginning of the 17th century.[7]

To search genealogy data and access online archives in Italy, you can start with the official Ancestors Portal provided by the *Ministero della Cultura*, located at https://antenati.cultura.gov.it. The website can be accessed in several languages, including English, French and German.

---

[7] Alain Nedjar, Liliane Nedjar, Gilles Boulu, and Raphaël Attias have created an impressive compilation of all the *ketoubot* (Jewish marriage contracts) from Livorno, which can be obtained here: https://bit.ly/ketoubot-livourne

Databases

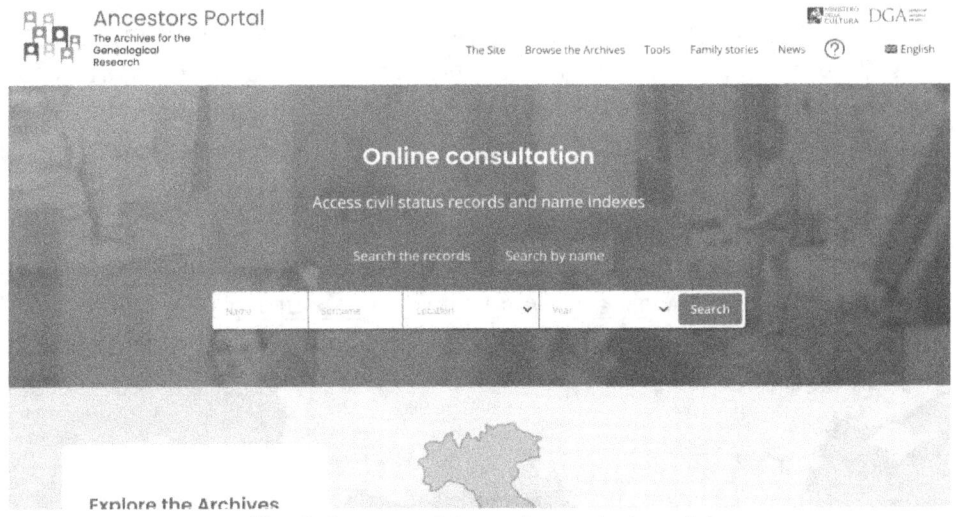

The Italian genealogy and ancestor's portal

The website provides a map of all online databases connected to the website, that you can access through a common interface.

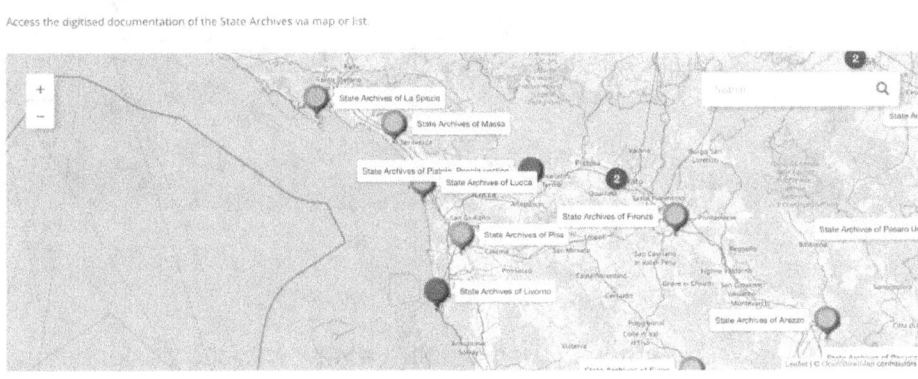

Just select a registry

4. France

The French national archives website is located at the following url: https://www.archives-nationales.culture.gouv.fr/. You can start your research process from its homepage, but I strongly recommend initiating specific searches from subsections, dedicated to specific topics.

For apprentice genealogists, the best thing to do is to start with the naturalization files. You can find them in the dedicated website section located at https://www.siv.archives-nationales.culture.gouv.fr/. Select the 'Naturalization and nationality' subsection, then enter a name and browse through all the results.

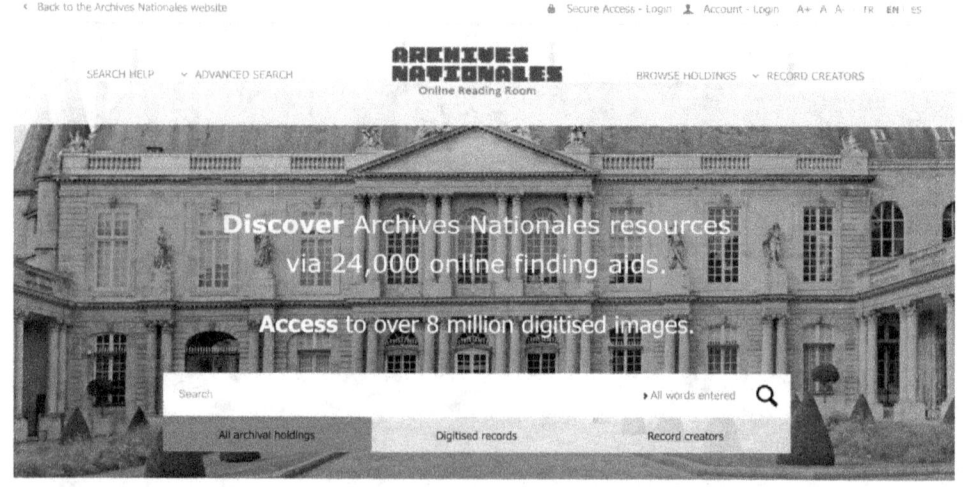

French "Archives nationales"

The French government launched in 2020 a new website designed as a main entrance to browse various databases for genealogy needs. Located at https://www.culture.fr/Genealogie, this very modern website still needs several improvements, the results it usually provides are too scarce to serve individual needs. You may prefer using the Gallica platform (see section G.2) for an extensive search over large sets of data.

Databases

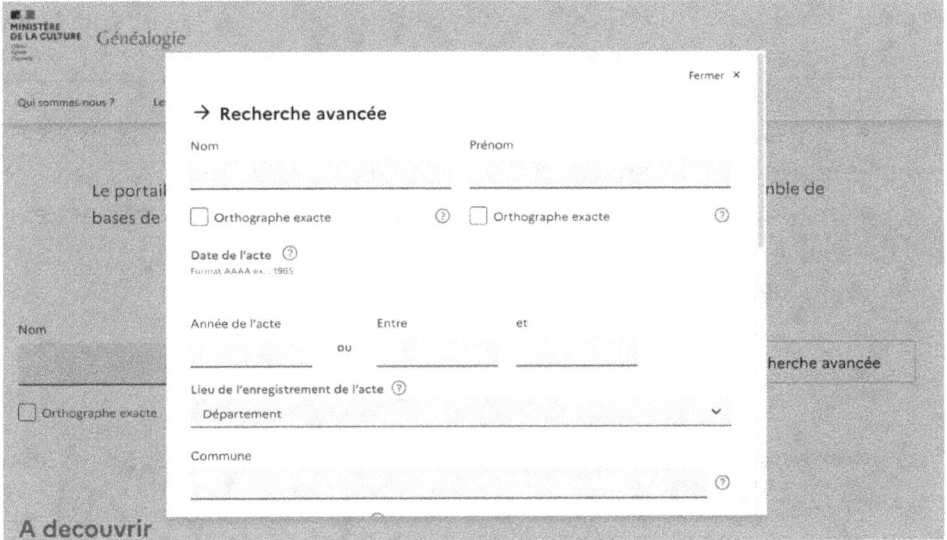

French government's official genealogy portal

5. Australia

There is no single official website for online genealogy archives in Australia. Civil registry databases are maintained by each Australian state and territory, making genealogical research a little bit tricky. Australian government birth, death and marriage (BDM) records are indexed and can be searched by name, place or date, within date ranges which are open to public searching access.

The National Library of Australia's website maintains a page that links to all state and territory BDM online registries at the following address : https://www.library.gov.au/research/family-history/births-deaths-and-marriages.

Please note that death certificates can only be requested by relatives or people mentioned in the registry. If you are not in this case, you must prove that some relative provided the right to order the certificate.

Besides the civil registry websites, you can find additional interesting data using the Record Search form included in the National Archives of Australia (https://www.naa.gov.au/). It can browse through several databases, including immigration and naturalization records, war and defence records, and copyright, patents and trademarks.

Search the National Archives of Australia

6. Canada

The Canadian national archives website is located at the following url: https://library-archives.canada.ca/. It provides various kind of data and a very friendly interface to census records from the website section located at https://recherche-collection-search.bac-lac.gc.ca/eng/census/index.

Databases

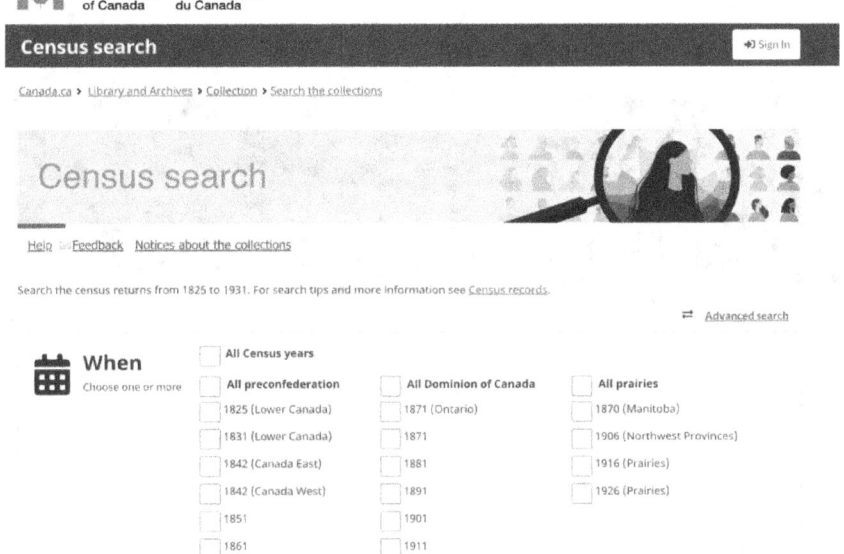

7. Germany

The German national archives website is located at the following url: https://www.bundesarchiv.de/. It provides a link to the archive portal located at: https://www.archivportal-d.de/. They include state archives, local archives, church archives and archives from universities and scientific institutions.

C. Official Gazettes

Official gazettes are government publications that provide public access to official notices, legal announcements, and legislative updates. They typically include laws, decrees, regulations, and administrative decisions, as well as announcements of public interest such as court decisions, tenders, and public appointments. They often contain a lot of useful data related to genealogy research. Some countries provide an open access to official gazettes archives, whereas others require a paid access.

For example, the Tunisian official gazette, whose archives are located at http://www.iort.gov.tn/ provide records of all revocations of citizenship since the mid 50s.

D. Municipal Archives

While national archives usually store data and records related to the national identity like naturalization files, local data information usually are stored at a more local level. That's why you should search municipal archives to find online records of birth or death certificates. Large cities

Databases

maintain a section in their municipality website dedicated to such requests, specifically designed for genealogists.

Here is a list of such municipal archives websites to get an idea of the kind of data you can find, and the way they are stored. Please note that, like for national data, there is no single philosophy or platform for such database design. So be prepared to some adventure…

1. New-York

The New-York City website possesses a genealogy section located at https://www.nyc.gov/site/records/historical-records/genealogy.page. It provides a very simple form to search individual birth, death, or wedding by last name.

The NYC Historical Vital Records Project

The New York City Municipal Archives is undertaking a mass digitization project to provide online access to 13.3 million historical birth, death, and marriage records.

Digitization Progress

77% (10,177,008 of 13.3 million records)

Check our progress

2. Paris

Paris archives can be accessed from https://archives.paris.fr/. You can browse several kinds of data, including city maps or cemetery registries online, and access civil registry databases for events that occurred after 1860 and before specific dates, to preserve people privacy: birth certificates until 1924, wedding certificates until 1948, death certificates until 1986, all limits active in 2024, that may change each year.

The way these certificates can be accessed is a little bit odd. The city of Paris maintains decennial tables with list of names born, married or dead in intervals of 10 years. They provide you with a record number that you must enter the specific form to request a display of 10 or 20 pages around the date of the event you're looking for. You then must browse the dates and record numbers until you find the one you're looking for.

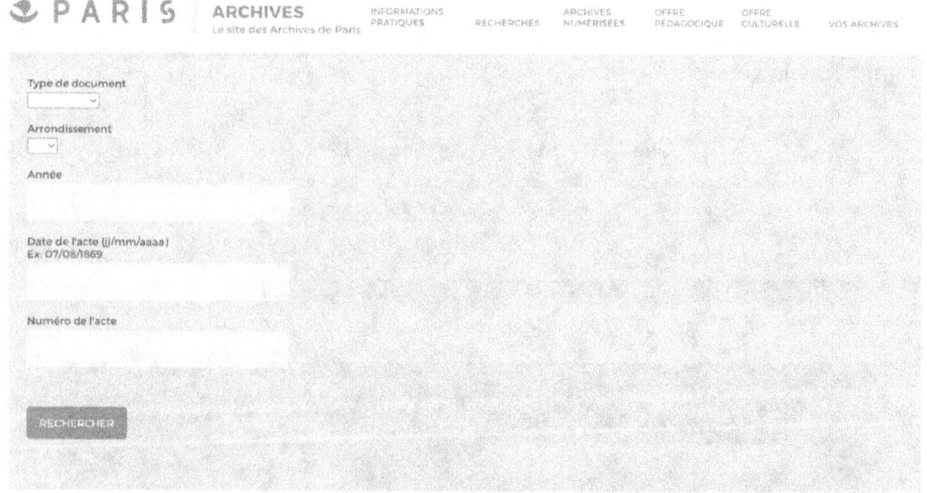

Paris is not the only French city to maintain online databases of civil registry records. Marseille, Lille, or Lyon also provide such services, although their websites interface and facilities differ from the Paris one.

Keep in mind that access to death certificate archives in France is subject to a date restriction. Death certificates less than 30 years old can still be ordered free of charge through the official French administration website

Databases

(www.service-public.fr) for most French cities, or via each city's respective website, which may have a different user interface.

E. Archives from Religious Institutions

Before the creation of modern states, and the organization of national census and civil registries, all civil data were stored by religious institutions. These records were maintained locally, in places such as churches. These documents were used to store various aspects of religious and community life. They often include records of baptisms, marriages, burials, and other significant events recorded by the church. Ecclesiastical archives are valuable for genealogical research, especially for periods or regions where civil records may be incomplete or missing.

The process of digitization of this kind of information is neither centralized, nor organized at the state level. Hence, you must make your own online research to check whether you can get access.

F. Familysearch

We have already briefly mentioned the existence of the FamilySearch website (www.familysearch.com) in the section dedicated to data exchange formats, specifically the GEDCOM format. FamilySearch is a site provided free of charge for genealogical research by the Mormon community, or more precisely by the Genealogical Society of Utah, the home base for the majority of the followers of this Church.

This site has several interesting aspects for online genealogical research. First, the longstanding nature of the initiative: the collection and preservation of genealogical documents stem indirectly from the obligation placed on the Church's followers, who venerate the eternal nature of families, provided their ancestors have been baptized. It is this injunction that gave rise to this immense undertaking, which has allowed the collection of millions of documents, many of which are in the form of microfilms stored in the Granite Mountain vaults, near Salt Lake City. Protected from unauthorized visitors for security reasons that are entirely understandable, these documents are freely accessible through the portal website www.familysearch.com.

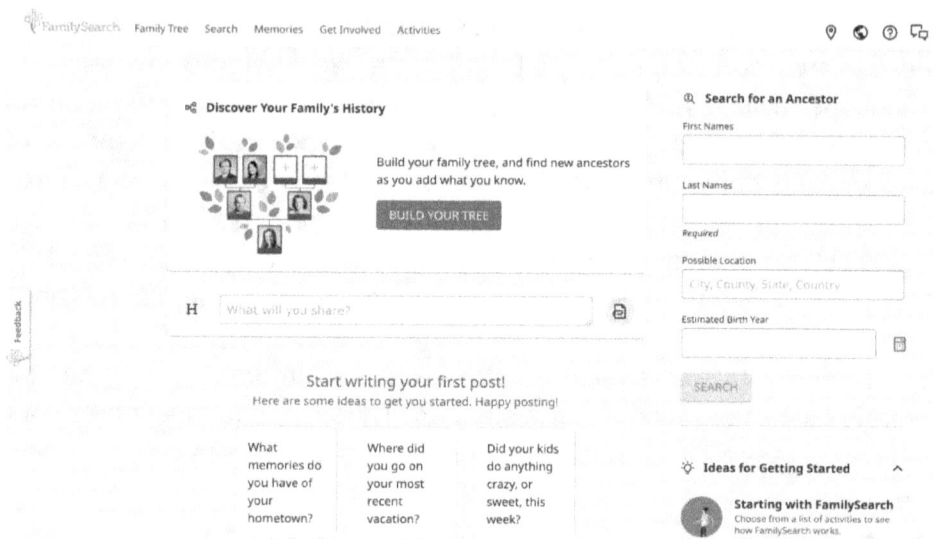

Familysearch very intuitive homepage

The second interesting feature is the diversity of sources compiled on FamilySearch. The site has established agreements with the national archives of several dozen countries. This allows users to conduct research across multiple geographic origins from a single platform, with a uniform structure, eliminating the need to master the specificities of regional online archive websites. Of course, these agreements are made under certain conditions to ensure that FamilySearch does not absorb entire national archives. For example, the agreement regarding civil records in France stipulates that the relevant registers must be at least one hundred years old.

FamilySearch also allows users to build their family tree for free directly on the site. While the provided interface is smooth and user-friendly, its range of features does not quite reach the level of tools like MyHeritage.

Finally, FamilySearch invites you to contribute to the global effort to continually enrich and improve the quality of the data made available. This can include participating in the indexing of new documents or validating existing data. Needless to say, you will not be compensated for the time you spend serving the community…

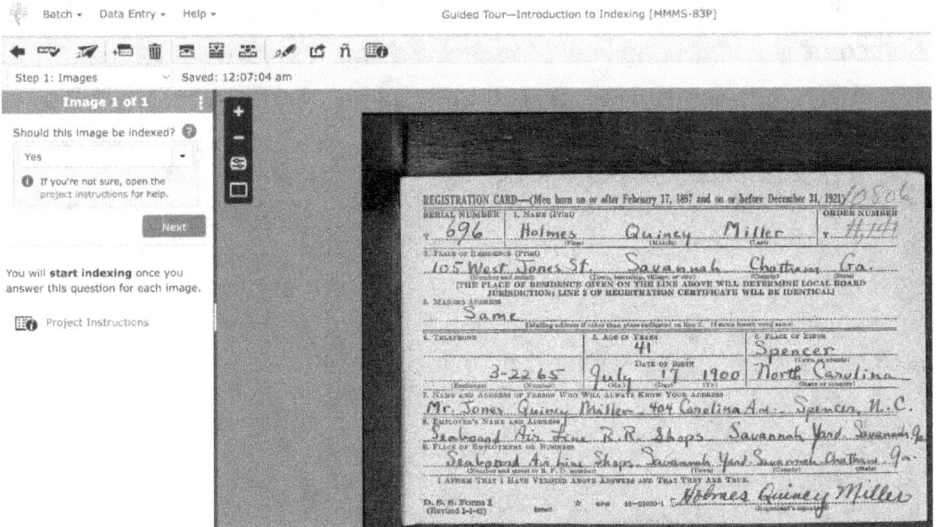

Everyone can participate to Familysearch efforts.

G. Specialized Websites

Some websites offer access to very specialized databases that could be useful for genealogists searching for specific names or events. Here are a couple of such websites.

1. Ellis Island

The Ellis Island website (https://www.statueofliberty.org/), managed by The Statue of Liberty-Ellis Island Foundation, is a comprehensive online portal that allows users to explore the rich history of immigration through Ellis Island. The site provides digital access to millions of historical immigration records, making it a valuable resource for those researching family history, genealogy, or U.S. immigration history.

Through the website, users can search passenger records for individuals who passed through Ellis Island and the Port of New York between 1820 and 1957. This includes access to names, ages, countries of origin, ship names, and arrival dates. To begin a search, users simply enter basic details like a name or an estimated year of arrival into the search fields. The site offers advanced filters to narrow down results by ship, departure port, or specific years, helping to pinpoint records even more precisely.

Once a match is found, users can view digital images of the original ship manifests, which provide fascinating insights into the lives of immigrants at the time. Additionally, the website includes tools for building a family tree, saving records, and accessing educational resources. It's a powerful tool for uncovering family stories and understanding the journeys of immigrants who helped shape the United States.

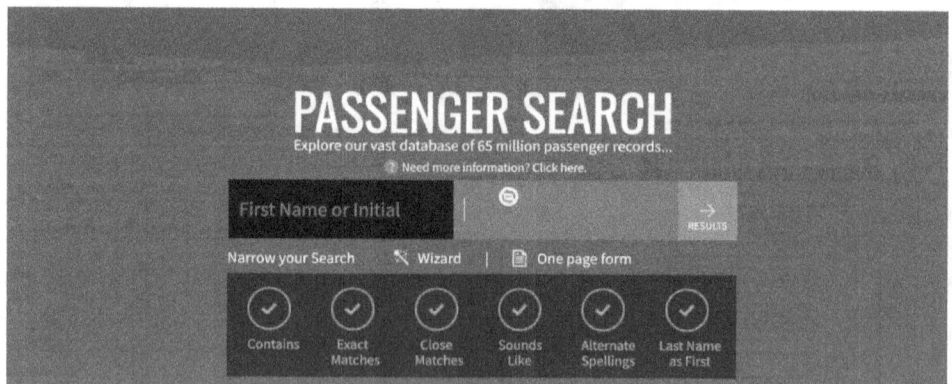

Ellis Island search form

2. Gallica

Gallica (https://gallica.bnf.fr/) is the website of the National Library of France. Since the late 1990s, this site has provided free access to millions of documents. A search engine, which can be challenging to master, allows users to find all occurrences of a given term: a family name, a city name, etc.

It would be impossible to cover all the features of a site as rich as Gallica here, as it would deserve an entire book on its own. However, for genealogical research, I highly recommend exploring the press archives, which are organized by themes, frequency of publication (daily, weekly, monthly), or by region or country of distribution.

These archives include documents in French as well as in foreign languages (English, German). Each document that positively matches a search on a surname can then be viewed with a viewer, page by page.

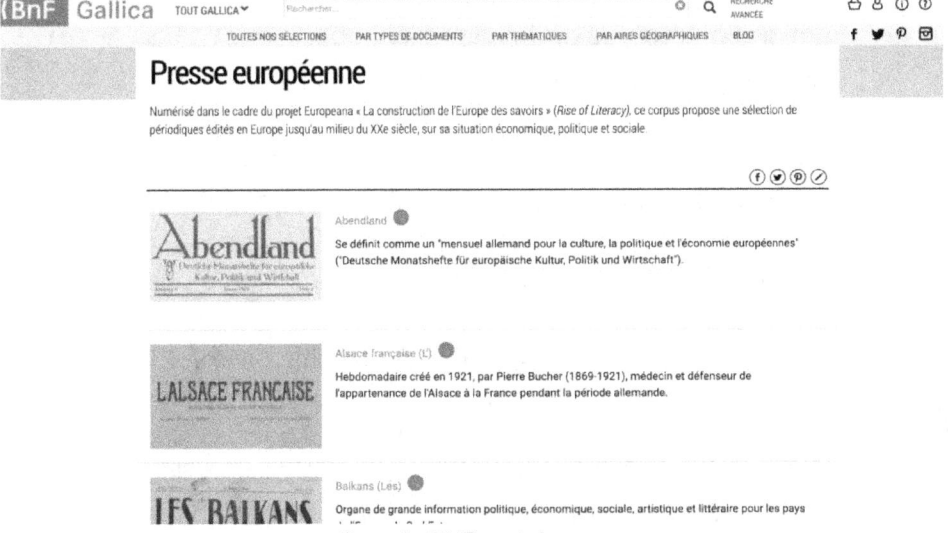

Sample BNF portal page.

H. Online Genealogy Software

To attract users and encourage them to subscribe to a paid plan, online genealogy software platforms offer access to additional databases that aggregate data from existing databases.

Take, for example, the site MyHeritage, mentioned earlier. This site offers searches across many external databases, some more complete than others, and the list of databases continues to grow over time. At the time of writing, this site offers, among other things, matches with:

- lists of celebrities
- lists of inventors
- lists of alumni from American universities
- public directories (notably in the United States)
- civil records, including divorce registers and church registers.
- lists of authors of books or scientific publications
- concentration camp registers
- immigration records and nationality applications
- passenger records at major American ports
- property deeds or criminal records (depending on the country)
- American social security applications and complaints
- archives from old newspapers published in the late 19th and early 20th centuries.
- and even lists of... registered chess competition players!

As you can see, this list of databases is extremely diverse. Some might criticize it for a lack of coherence, perhaps. But in practice, it must be acknowledged that it is a true powerhouse, because as soon as MyHeritage signs an agreement with a new database holder, its users gain immediate access to millions of original documents, remarkably indexed by name - even though some errors persist.

I. Facebook

Facebook is, of course, not a genealogy software. But the information found there can be useful for retrieving genealogical data, although it shouldn't be expected to trace back through many generations.

1. Searching for Lateral Information

Personal information, such as date or place of birth, is sometimes mentioned on users' profiles. However, be cautious in trusting this information, as profiles are not always filled out carefully, and some users may include false information to mislead potential snoops.

Family relationships are also sometimes mentioned on users' profiles: father, mother, spouse, son, daughter, sister, brother, cousin, uncle, aunt, grandfather, grandmother, grandson, or granddaughter—every kind of relationship is available. Again, stay vigilant, as some people label their brothers-in-law or sisters-in-law, or simply close friends, as "brother" or "sister," a trend especially common among younger generations.

Occasionally, users, not fully understanding how Facebook works, may reverse relationships and call themselves grandsons or granddaughters when they are actually grandparents.

Finally, browsing public posts, you may sometimes find information, although it should be interpreted with caution: thanks exchanged between siblings or cousins, often on birthdays.

Some people tend to commemorate the death anniversaries of their parents or grandparents, which can help fill in missing information. Others share photos of their parents or grandparents and sometimes even gravestone photographs, providing dates or ages.

In short, while Facebook doesn't allow for structured research, it can occasionally help complete a family tree with information that would otherwise be hard to find.

2. Genealogy Research Groups

Many amateur genealogists gather in Facebook groups dedicated to searching for and sharing information to complete family trees. Several types of groups exist.

In surname groups, you'll find people with the same or similar last names who are trying to trace their origins or familial connections. These groups may have several hundred members, depending on the popularity of the name.

In geographic origin groups, members share a common origin, such as a country or region, and try to find connections between different families. Such groups often have thousands of members.

If you don't find a group for your surname or geographic origin, nothing prevents you from creating your own. It's very simple, and Facebook even provides a guide at https://www.facebook.com/help/167970719931213.

You will then need to maintain the group by creating some initial content, like a tree excerpt or original documents, and inviting members, encouraging them to invite others to join the group.

You'll also need to moderate access and content to ensure the group retains its identity and editorial line, so new members aren't driven away by content that diverges from what they expected to find.

Such a group can only function if a significant percentage of its members participate. To encourage this, you could, for instance, recognize new members each week and encourage them to share their family tree or ask unique questions. Facebook offers a feature that allows group administrators to manage this type of engagement easily.

V. Introducing DNA

A scientific revolution has significantly impacted genealogical research in recent years: the sequencing of human DNA. Let's examine its consequences together.

A. Principles of DNA

In 1953, two researchers, American James Watson and British Francis Crick[8], discovered the double-helix structure of a macromolecule that would revolutionize medical research: DNA.

This giant molecule is the common building block of all living organisms on earth with cells that have a nucleus: the eukaryotes. This includes a wide variety of animals and plants, from earthworms to chameleons, from date palms to whales. Each cell in a human body has a nucleus that contains nearly all of its DNA (except for mitochondrial DNA, which is housed elsewhere within the cell).

Each strand of this helix contains thousands of amino acids, or bases, which can only be one of the following four types: adenine, cytosine, guanine, and thymine. The number and sequencing of these bases determine both the species of the organism being studied and its genes. DNA thus serves as both an identification card for the species (by number and arrangement) and the individual (by specific sequences at predetermined locations). This is known as a "karyotype."

Understanding this sequencing was a crucial goal for many scientific teams, but the appropriate technology was needed to achieve it. It would take another fifty years before the sequencing of human DNA was finally

---

[8] For a more detailed description of the discovery of DNA, see *Genome Hacking* by Deborah Levy

Introducing DNA

completed in 2003, meaning scientists understood the distribution of genes on a strand containing 3 billion nucleotides.

But that's not all. Human DNA is distributed across 23 chromosomes, or rather, pairs of chromosomes. Every human inherits one strand of each pair from each parent, though not quite entirely: one of the chromosomes is responsible for sexual differentiation, so males have a distinct X and Y pair, while females have an identical X pair. In this way, we inherit half of our genetic heritage from each parent.

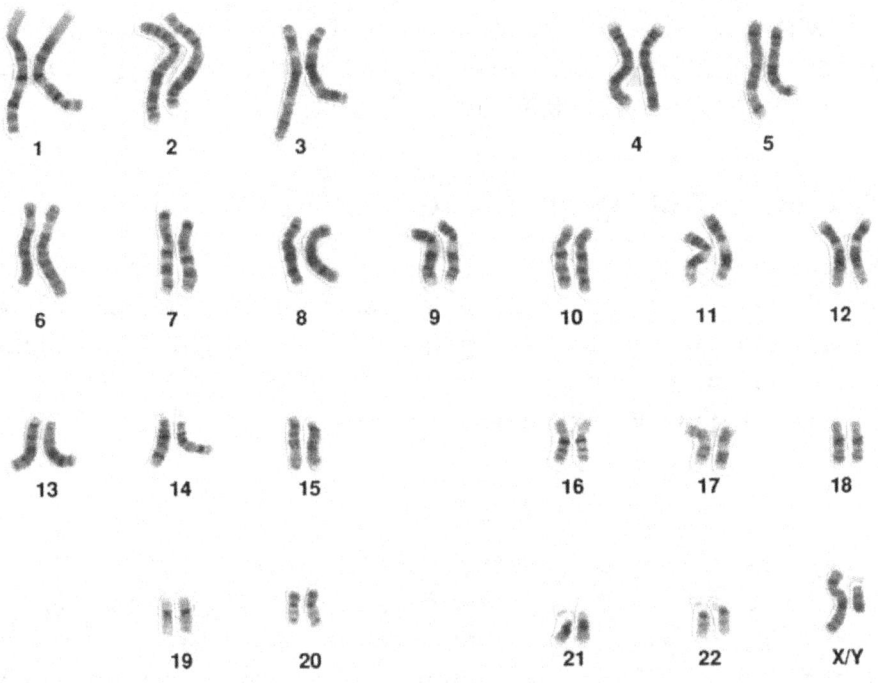

Karyotype of a male individual (XY)
Source: Wikipedia / National Human Genome Research Institute

If we go into detail, things are a bit more complex. We mentioned earlier that the number and sequencing of the double helix bases determine the species. This is why we humans have two arms, two eyes, two ears, skin, a heart, intestines, etc. This does not vary from one individual to another. What varies are the characteristics of these organs: eye colour, ear shape, arm length.

This uniformity coupled with this great variety comes from the fact that two human beings share a large portion of their DNA, common to the entire species, which represents about 99.7% of the genome. This part is called non-coding DNA. The remaining 0.3% serves to differentiate individuals.

Thus, when we inherit half of our genetic material from each parent, it means we inherit half of the coding part: the non-coding part is common to both our parents, regardless of their origin.

B. Sequencing and Interpretation

We mentioned earlier that DNA has brought a real revolution to medical research. This is also true for genealogical research, though for different reasons. Knowledge of the coding part of DNA between two individuals allows us to determine the percentage of (coding) DNA in common and infer a kind of genealogical distance.

1. Parents and Children

The logic is simple. You possess 50%[9] of each of your parents' genetic material, and your children possess 50% of yours. They are, a priori, the only people with whom you share 50% of your DNA. Suppose, for instance, you don't know your parents but meet an older person who shares 50% of your DNA: this person is one of your parents. And if it's a younger person, they are one of your children. You can imagine the potential of this approach for paternity research.

---

[9] Approximately 50%, though there are always small variations.

But that's not all. Since you share 50% of your DNA with each of your parents, who in turn share 50% with theirs, it means you share 25% with each of your four grandparents—assuming your parents are not cousins, as that could complicate matters.

This logic applies upward but also downward: you thus share 25% of your DNA with your grandchildren. The comment we made earlier on paternity research enables us to identify family ties with 25% shared DNA, or, by repeating the operation with great-grandparents or great-grandchildren, with 12.5% shared DNA.

2. Siblings

Things get more complicated with other family members, as a degree of randomness comes into play. For example, consider two siblings, regardless of gender. Each child shares 50% of their DNA with their common parents, but how much DNA do they share with each other?

Setting aside the case of identical twins—who share the same genetic material and thus 100% of their DNA—in the case of each chromosome, each child inherits half of the parents' genetic material. This half could be the same, the complementary half, or often several similar fragments. This process occurs across each of the 23 chromosome pairs, making the probability of a complete difference (0%) very low, as well as the probability of a perfect identity (100%). Statistically, two siblings share approximately 50% of their genetic material on average, even if they may have 42% or 56%.

Consequently, if you meet someone who shares about 50% of your DNA, it is either one of your parents, one of your children, or one of your siblings. Age is the main distinguishing factor here, although in large families, siblings may have significant age differences.

The same logic applies to nephews and nieces (on average 25% shared DNA) or aunts and uncles (on average 12.5% shared DNA).

3. Half-Siblings

Half-sibling relationships are not uncommon. The increase in blended families is one contributing factor today. But a century ago, another factor was significant: the high maternal mortality rate during childbirth. It was common for a widower to remarry and start a second family—bringing with it inheritance issues that have fueled 19th-century literature.

What percentage of DNA do half-siblings share? The logic remains the same as before: each inherits genetic material from one shared parent, so they share 25% of their DNA on average.

4. Consanguineous Marriages

Though becoming increasingly rare, consanguineous marriages have always existed for various reasons, including preserving family assets, customs, cultures, or material wealth.

Today, we know such unions have not always been without consequences, sometimes severe, for the health of descendants. The reason is that, in cases where a mutation causing a rare genetic disease is common in both parents, their children are at risk of developing a condition that would not have appeared if one parent did not carry the mutation. With a non-carrier partner, the mutation would have been passed down without risk.

In terms of the percentage of shared DNA among family members, consanguineous marriages disrupt statistical norms.

5. Recap Table

Based on the percentages mentioned above, we can infer the following table, which can serve as a reference for analyzing results from a DNA kit comparison between two individuals.

Shared Percentage	Younger Person	Same Age Person	Older Person
100%		Twin	
~50%	Child	Sibling	Parent
~25%	Grandchild	Uncle/Aunt	Grandparent, Uncle/Aunt
~12%	Great-grand Child, Nephew/Niece	First cousin	Great-grand-parent
~6%	Child of First Cousin		Parent's First Cousin
~3%	Grandchild of Parent's Cousin	First Cousin Once Removed	Grandparent's First Cousin
~1,5%	Child of First Cousin Once Removed		

C. Practical Implementation

To use DNA in genealogy work is quite simple. Just visit a website that offers DNA testing kits and place an order for one or more kits, especially if you're trying to trace the ancestry of multiple people. For example, you can visit the websites of companies such as 23andme

Introducing DNA

(www.23andme.com) or MyHeritage (www.myheritage.fr) to compare their offerings.

These kits typically cost between 40 and 100 euros, depending on the provider and the time of year (promotions are common around Black Friday or special dates like Valentine's Day, Mother's Day, etc.). Note that some countries, like France, prohibit private DNA sequencing, and violations may result in fines. Although sanctions are relatively rare, these kits can no longer be shipped to restricted countries as of recent years.

Once you've ordered, it will take a few days to receive your kit, which usually includes two vials and two swabs (the double set helps reduce the risk of errors). Swab the inside of your cheek, place it in a vial, put the vials in the provided envelope, and mail it. You'll need to wait another three or four weeks to receive the results, which you can check on the company's website.

The results usually include several sections. The first part shows your likely origin: European, Asian, North African, or South American? More specifically, you'll get a breakdown by percentage, such as 50% Spanish, 30% Chinese, and 20% African. While these insights are interesting, they should be taken with a grain of salt as they are only comparisons against the site's DNA database and reflect relative origins rather than absolute ones.

You'll also discover your haplogroup, which represents a group of people with shared genetic characteristics due to a common ancestor. Haplogroups are organized hierarchically, with each human group descending from a larger ancestral group that existed thousands of years ago.

Another significant part of the results allows you to compare your DNA with other users on the platform. This is particularly valuable for genealogy, as you'll receive a list of several dozen users with the percentage of DNA you share, which can be a reliable indicator to identify first or second cousins.

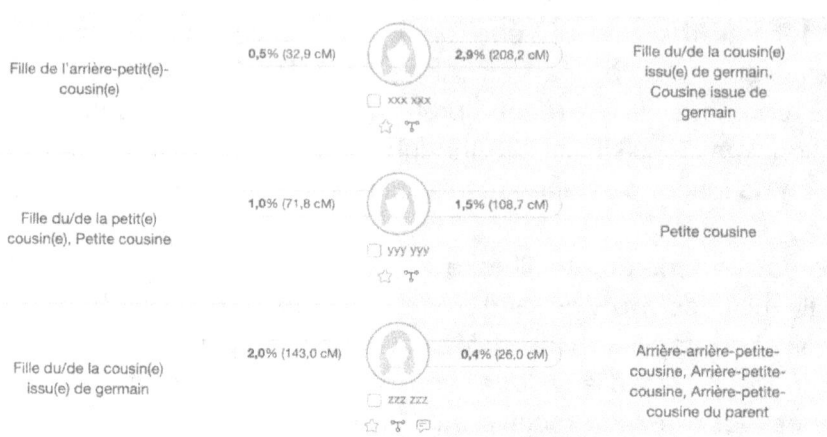

Fille de l'arrière-petit(e)-cousin(e) | 0,5% (32,9 cM) | xxx xxx | 2,9% (208,2 cM) | Fille du/de la cousin(e) issu(e) de germain, Cousine issue de germain

Fille du/de la petit(e) cousin(e), Petite cousine | 1,0% (71,8 cM) | yyy yyy | 1,5% (108,7 cM) | Petite cousine

Fille du/de la cousin(e) issu(e) de germain | 2,0% (143,0 cM) | zzz zzz | 0,4% (26,0 cM) | Arrière-arrière-petite-cousine, Arrière-petite-cousine, Arrière-petite-cousine du parent

Example of DNA match result on MyHeritage

From a technical perspective, the result of your DNA test is actually a digital file that you can download and store at home. You can even use the same file across multiple platforms. For example, you might use 23andMe for your test, then download the file to upload it to MyHeritage to find relatives on that platform.

The DNA sequencing file can also be used on sites that allow DNA analysis for various profiles, like GEDmatch (www.gedmatch.org). This company, founded in 2010 by Curtis Rogers and John Olson, differs from traditional genealogy sites and software. The original purpose of GEDmatch was to allow anyone with a DNA sequence to compare it with a database of other sequences uploaded by millions of users, helping to detect common ancestors and thus identify distant relatives based on the percentage of shared DNA. GEDmatch accepts files from 23andMe, MyHeritage, Ancestry, and other companies.

With an interface that may seem somewhat complex for beginners, GEDmatch enables advanced comparisons, including the ability to detect if your parents share a significant amount of coding DNA, indicating common ancestors.

D. Y Chromosome and Mitochondrial DNA

As previously mentioned, an individual's DNA is inherited approximately equally from their father and mother, with a few specific details. These details are discussed in the following paragraphs.

a) Y Chromosome

Let's start with the Y chromosome. At the beginning of this chapter, we noted that each pair of chromosomes comprises 50% of the genetic material from the corresponding chromosome of the father and the mother. This is true for 22 pairs of chromosomes but not for the 23rd pair, which determines the individual's sex. This chromosome comes in two forms: X and Y. Women have a double XX pair, while men have an XY pair. Specifically, women inherit an X strand from each parent, while men inherit the entire Y strand from their father.

This characteristic has an obvious consequence: all male individuals from the same lineage share the same Y chromosome, except for minor mutations. Thus, by comparing Y chromosome lineages, it is possible to determine whether individuals have the same male ancestor.

This is the principle followed by FamilyTreeDNA, a company offering a DNA kit and Y chromosome comparison with other users (restricted to men, naturally). Additionally, this site includes a database of DNA from individuals found at archaeological sites, allowing customers to trace the origins of their most distant ancestors across several centuries.

b) Mitochondrial DNA

As noted earlier, a small portion of DNA does not reside in the cell nucleus but within the cell itself: mitochondrial DNA[10], sometimes referred to as mtDNA. This molecule is distinct from the DNA contained within the cell nucleus, both in its structure and in the number of bases it contains. Another unique aspect is that mitochondrial DNA is entirely inherited from the mother.

Thus, by reasoning in the same way as with the Y chromosome, it is possible to trace maternal lineages by comparing only mitochondrial DNA. This is also a service offered by the site FamilyTreeDNA.

E. Ancient DNA

In 2022, the Nobel Prize in Medicine was awarded to Svante Pääbo. The son of a Nobel laureate in medicine[11], this Swedish researcher based in Germany pioneered a revolutionary technique that allows for the analysis of DNA extracted from tiny fragments of a human or animal, even if the DNA comes from an individual who died thousands of years ago — commonly referred to as ancient DNA.

The work of Svante Pääbo demonstrated, for instance, that a portion of our genetic heritage comes not from Homo sapiens, but from Homo neanderthalensis, thus proving that the two species were sexually compatible...

As scientific interest in ancient DNA has grown, some websites have begun to commercially exploit this information, offering users the chance to compare their DNA, sequenced via a kit like those offered by 23andme or MyHeritage, with DNA extracted from remains found at archaeological sites.

---

[10] Mitochondria serve as an energy reservoir for the proper functioning of cells.
[11] His father, Sune Bergström, received it in 1982!

Such is the case with My True Ancestry (www.mytrueancestry.com), a company that, using a DNA sequence from one of the more common providers, offers a service for a flat fee allowing users to compare their DNA with dozens of sources, identifying the diverse ancestries that make up their genetic heritage among 150 ancient civilizations, ranging from Aboriginals to Vikings...

This type of information thus helps confirm a more or less ancient origin, though with a relatively low degree of certainty and precision. However, the relevance is quite limited when it comes to the construction of a genealogical tree...

F. A Barrier: Legislation

While DNA sequencing undeniably offers advantages for genealogical research, it has become increasingly difficult for French residents to benefit from it. The reason is simple: DNA sequencing is only authorized in specific contexts (for medical or judicial purposes) and is prohibited for personal use. Anyone violating this restriction may face a fine of up to 3,750 euros.

In reality, although the tests are prohibited, few people have actually been fined. Until a few years ago, it was still possible to receive DNA sequencing kits from companies like MyHeritage or 23andme. Unfortunately, this has been impossible since 2022 in countries like France, likely because these companies themselves faced the threat of heavy penalties.

France is not the only country where it has become impossible to receive a DNA kit. Individuals in Germany, China, and Russia face similar difficulties, whether in democratic regimes or more authoritarian ones.

Why has it become increasingly difficult, if not impossible, to access this technological advancement? There are several possible reasons. The first cited reason is consumer protection. Implicitly, this suggests that a person taking a DNA test may not be fully aware of the potential implications, such as discovering an unknown part of their family. This is a somewhat valid reason: by taking such a test, the consumer explicitly expresses interest in the resulting information.

The second commonly cited reason is the protection of personal data. Since DNA is equivalent to a digital ID, would you be willing to leave a copy of such a document with a company you know nothing about? This argument is valid, and the data breach at 23andme a few years ago[12] supports the authorities' efforts to protect their citizens. However, this is a conscious risk. After all, you regularly entrust private data to other companies susceptible to hacking. Why should the state intervene to manage data security between companies and their customers?

The third reason is even more subtle. It involves the idea that by undergoing sequencing with foreign companies, you allow that country to profile the citizens of each nation. This type of information, while merely statistical today, could become far more dangerous if an entity were able to develop a biological weapon targeting individuals with specific genetic markers. Is this more science-fiction than reality? Technological progress has been so rapid in recent decades that a touch of paranoia in national governance may not be...

---

[12] Cf. https://bit.ly/23andmehacked

VI. Final Recommendations Before You Begin

As you can see, building a family tree isn't that difficult, whether done by hand or with software.

However, before embarking on such a project, here are a few final recommendations to keep in mind and review over time, as we all tend to overlook security or best practices once we feel comfortable with a tool of any kind.

1. Be Precise

The quality of a family tree directly depends on the quality of the information entered. If mistakes creep into your tree, correct them as soon as possible, as you may forget the exact values and risk spreading false information to others.

Pay close attention to the spelling of names and surnames you enter, as well as dates and places of birth or death. When translating from a non-Latin alphabet to a Latin one, for example from Russian or Arabic to French, you may choose to keep both forms.

2. Be Thorough

Whenever possible, try to enter as much information as you can about an individual at the same time. From experience, it is indeed more challenging to revisit information that was already entered in the past. Therefore, include all available data: all children, all aunts and uncles, all dates.

Additionally, when sharing your family tree with others, consider the psychological impact this information might convey. A missing individual could be perceived very negatively by their relatives.

3. Cite Your Sources

When mentioning events from the distant past, always indicate the source of the information whenever possible: the online database, the family tree, or if the details came from an interview with a descendant of the person in question, etc. Whenever possible, keep a digital copy of the documents you used.

Keep these sources up to date over time, as it may become very difficult, or even impossible, to trace them back years later when someone asks where you got a particular piece of information.

4. Verify Your Sources

Citing your sources is one thing, but verifying them is another. Don't hesitate to question the reliability of the data you rely on.

For example, if you add information from another family tree, check the sources that the tree's creator used. They may have been less cautious than you, and the data you find could be degraded or even incorrect.

Similarly, if you interview an elder, which is a great idea, remember that their memory may not be flawless. It's easy to recall the names and details of parents, siblings, their spouses, and children. But over time, the family expands, and remembering the names of grandnieces, or the siblings of one's grandparents, can become challenging...

5. Make Backups

If you're using genealogy software, make regular backups to protect yourself against any incidents. If it's software on your computer, you'll likely be able to export in GEDCOM format. Try to do this at least once a year, or at significant milestones, like reaching hundreds or thousands of individuals in your tree.

If it's online software, download a copy of your work to avoid losing everything if the site is hacked or shuts down.

6. Respect People Privacy

A genealogy software program isn't a personal notebook. Out of respect for individuals' privacy, avoid storing sensitive data, like physical addresses, building access codes, passwords, or any other information that could be exploited by malicious users.

Whenever possible, and especially if you're using online software, mask data about living individuals. Avoid storing photos of very young children.

And most importantly, only share access to your trees with trusted individuals.

7. Follow Team Rules

It is common not to be the only one working on a family tree. If that's the case, and you're working with others on the same tree, try to enforce the previous rules with the other team members.

Also, strive for consistency in naming conventions so the data remains uniform, regardless of who entered it. For example, if you decide to write surnames in uppercase, make sure everyone follows that. If you choose to list the city and country of an event separated by a comma, without mentioning the department or state, follow the same rule everywhere.

A Beginner's Guide to Genealogy 2.0

VII. Conclusion

Over the past fifteen years, due to the technological acceleration brought about by digital exchanges and the decreasing cost of DNA sequencing, genealogy has experienced a new boost. A few clicks are all it takes to create one's family tree and, through the magic of "matching" on online genealogy software, reconnect with entire branches of one's family that had been lost from view.

Genealogical research is no longer the exclusive domain of dusty experts charging unusually high fees for their services. Anyone can start a genealogical research project for free or with just a few euros. Millions of people have tried these tools, managing to restore ties that history or geography had loosened and to bring distant cousins together.

Some may see it as nothing more than a trivial pastime, as anecdotal and sterile as using social media. A fad that will eventually pass. Others may view it as the harmful effects of rampant globalization and a dangerous challenge to the right to be forgotten, due to the risks it poses to family integrity through paternity searches.

The continuous renewal of generations and the unquenchable need to rediscover our origins will dismiss supporters of the first view. As for the second, it's up to both lawmakers and the owners of online platforms and software to set limits on usage and to curb the potential misuse of these tools.

In the end, I see it as just one more tool to remind us that, despite our diversity, antagonisms, or conflicts, we are all part of one and the same family: the human species.

A Beginner's Guide to Genealogy 2.0

VIII. Acknowledgements

A book is often the product of an author's work, or a small team of co-authors. But in reality, it could not come to life without the help of countless implicit contributors, who have contributed, sometimes unknowingly, to its creation.

I would therefore like to thank my "comrades" in genealogy, David Gamrasni first and foremost, but of course also David Liscia, Emmanuel Fitoussi, and Jean-Marc Benhamou for the magnificent work they have been doing for many years, as well as Nathan Hattab, Elena Gihan, Moshe Uzan, and many others, who have helped me in countless ways to build the immense tree I have been working on for the past ten years. Thanks to Liliane Nedjar, Alain Nedjar, Thierry Samama, and Gilles Boulu, whose work on the CGJ's BECANE database has greatly facilitated my task. Thanks to the thousands of amateur genealogists whose work I have relied on over the years.

Thank you to the dozens of unknown (or known) computer scientists who enabled the rise of the Internet and online genealogy sites, my nearly daily working tools.

Thank you to the thousands of civil servants who have tirelessly worked over the decades to build this immense memory of humanity, our civil records.

Finally, thanks to my wife and children, who have shown boundless patience in recent years, accepting that I sometimes spend more time deciphering birth records than with them.

A Beginner's Guide to Genealogy 2.0

IX. References

A. Bibliographic References

Cavalli-Sforza, Luca et Francesco, *The Great Human Diasporas: The History of Diversity and Evolution*, Addison-Wesley, 1995

Darlu, Pierre, *Origines : l'ADN a-t-il réponse à tout ?*, Le Pommier, 2016

De Morant, Guillaume, *Les mormons et la généalogie*, Archives & Culture, 2015

Grigorieff, Nathan, *Construire son arbre généalogique*, Eyrolles, 2004

Jovanovic-Floricourt, Nathalie, *L'ADN un outil généalogique*, Archives & Culture, 2019

Lévy, Deborah, *Genome Hacking*, Atlande, 2021

Quintana-Murci, Luis, *Human Peoples: On the Genetic Traces of Human Evolution, Migration and Adaptation,* Allen Lane, 2021

Reich, David, *Who we are and how we got there*, Pantheon, 2018

B. Online References

https://en.wikipedia.org/wiki/Comparison_of_genealogy_software

1. Genealogy Websites

- www.ancestry.com
- www.familysearch.com
- www.familytreedna.com
- www.filae.com
- www.gedmatch.com
- www.geneanet.fr
- www.geni.com
- www.myheritage.fr
- www.mytrueancestry.com

X. Glossary

23andme
23andme is an American company that was one of the first to offer DNA kits to the public.

Act
An official document recording a family event: birth, death, marriage, naturalization, etc.

Ancestral Tree
A family tree representing the lineage of ancestors from generation to generation.

centiMorgan
The centiMorgan (abbreviated: cM) is a measure of genetic distance used to quantify the distance between two genes on a chromosome.

Civil Status
The civil status of an individual represents their situation regarding family and society. It includes various acts (birth, marriage, or death certificates, naturalization decree) that precisely define this individual.

Collateral
A relative who is not in the direct line: sibling, cousin, uncle, aunt, etc.

Cousin Reunion
A gathering of cousins, more or less distant, from a common ancestor. By extension, a gathering of people with the same last name.

Degree of Kinship
The number of generations separating two individuals.

Descendant Tree

A family tree representing the lineage of descendants from generation to generation.

Descendants
The group of people descending from a common ancestor.

Digitized Archives
Archived documents made available in digital format, usually as an image viewable on a website.

DNA
DNA is a set of molecular structures containing genetic information inherited by everyone from their parents.

DNA Sequencing
DNA sequencing is the process of going from a cell sample to establishing a kind of genetic ID.

Family Tree
A graph of relationships between individuals, showing links between parents, children, and spouses.

First Cousin
Individuals who share a common grandfather or grandmother.

Facebook
Facebook is a social network created in 2004 by Mark Zuckerberg, where each member can share certain information useful for genealogy research: parents, children, birth dates, etc.

Filiation
The kinship link between children and parents.

Gallica

Glossary

Gallica is the site providing access to the digital archives of the National Library of France.

GEDCOM
GEDCOM is the name of the main data exchange format between genealogy software.

GEDmatch
GEDmatch is a site that allows users to compare a DNA kit with millions of other DNA kits to check for common ancestors.

Gene
A gene is a segment of DNA. It consists of a sequence of nucleotides that corresponds to a genetic sequence for synthesizing a specific protein.

Geni
Geni is an online genealogy software and site, created in 2006 and acquired by MyHeritage.

Haplogroup
A haplogroup is a group of people who share similar genetic characteristics from a common ancestor.

Kin
All relatives of a person.

Lineage
The lineage of an individual is the succession of their ancestors and descendants.

MyHeritage
MyHeritage is an online genealogy software and site that allows users to create their family tree, perform DNA testing, and find relatives sharing at least 0.1% of DNA.

Second Cousin
Individuals whose parents are first cousins.

Surname
Family name.

XI. Just one more thing…

If you enjoyed this book, and even if it did not meet your expectations, feel free to leave a comment on Amazon by <u>clicking on this link</u> (if you're reading the Kindle version) or by scanning this QR code.

www.ingramcontent.com/pod-product-compliance
Lightning Source LLC
Chambersburg PA
CBHW052333220526
45472CB00001B/408